Woodwork Technology

T. Pettit

Formerly Head of Creative Studies Faculty,
Aireville School, Skipton, North Yorkshire.

Edward Arnold

Contents

Introduction

The aim of this book is to provide all woodwork students, from non-exam students to those preparing for first examinations, with the essential and up-to-date information about the tools and materials used in the subject, in the most simple and direct way.

The book contains four main sections; tools, timber, veneers and manufactured boards, assembling and finishing. Each section has been broken down into concise sub-sections which summarise the key elements of that topic. Each sub-section is followed by simple and straight forward questions to check that students have understood the main teaching points. At the end of each main section there is a wide selection of revision questions for exam preparation and more open ended investigation.

The text is written in clear uncomplicated language avoiding unnecessary detail to give students the maximum practical help. Description and explanation in the text are supported by many photographs and clear line drawings to stimulate student interest. In this way the book can be used in class, or independently for revision and study.

Many of the questions have been devised to encourage students to be more observant and enquiring when working in a school workshop. It is assumed that students will draw upon this workshop experience and that the book will be a means of furthering a deeper appreciation of the subject.

Also available is a pack of 45 blackline masters by the same author, **Woodwork Technology: Construction Skills Pack**, which shows the essential construction techniques used in Woodwork. This plus the book provide a complete woodworking system.

Acknowledgements

The publishers and author would like to thank the following for permission to use copyright illustrations: Bahco Record Tools Ltd: pp 1, 2, 3 & 4 (photos only), 6, 7, 8 (except Fig. 32), 9 & 10 (photos only), 11, 12, 13 (photos only), 14 (except Fig. 63), 15 (except Fig. 68), 16, 17, 18, 19l, 22bl; Kango Wolf Power Tools Ltd: pp 19r, 20tl, 20bc, 20br; Boxford Ltd: p. 20bl; T S Harrison & Sons Ltd: pp 21l, 22 (Figs 106–8); The 600 Group PLC: p. 21cr; Museum of English Rural Life: p. 28tl; UAC International Ltd: p. 28cl; Ciba-Geigy PLC: pp 32, 39 (photos only) 49; Dalescraft Furniture Ltd: p. 41 (photo only); Formica Ltd: p. 42bl; GKN Screws & Fasteners Ltd: pp 46 (Figs 181 & 182), 48t; Woodfit Ltd: pp 47 (Figs 184 & 185), 51 (Figs 198 & 199), 52 (Figs 200 & 202), 53r, 54cl, 54bl, 55tl, 55tr, 55cr, 55br, 56 (except Fig 217); Universal Fittings (Kingsland) Ltd: p. 56br; Norton Abrasives Ltd: p. 57; E Parsons & Sons Ltd (Blackfriar Paints): pp 58, 59.

All other illustrations supplied by the author.

General Acknowledgements

The author would also like to thank the following for their advice and assistance: Borden (UK) Ltd, Spear & Jackson (Tools) Ltd, Chipboard Promotion Association, Alfred Green & Co (Skipton) Ltd, - a subsidiary of Southern-Evans Ltd, Finnish Plywood International, Neill Tools Ltd, The Design Council, The Rawplug Co Ltd, Worth Photofinishers Ltd, Lervad (UK) Ltd, English Abrasives Ltd, British Standards Institution, Stanley Tools.

Grateful appreciation is expressed to my daughter Angela who typed the manuscript.

Timber charts were typed by Mrs A Nicholson.

ONE

Tools

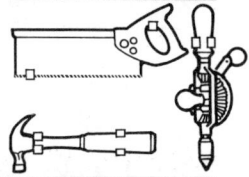

Many modern tools are of traditional design resembling those made thousands of years ago. Today modern steels are used in their manufacture to give longer and better service, and plastics are often used in place of wood for handles and some small components. If you expect to produce fine craftsmanship use only good quality tools which are well cared for.

1.1 MEASURING, SETTING OUT AND TESTING

A **soft pencil** is used for marking out on wood because the lead does not score the surface, and any marks can be easily removed. Measurements are taken with a **steel rule** (which is also used for testing for straightness), with a **folding rule**, or for longer measurements with a **steel tape** (Fig. 1).

The **try-square** (Fig. 2) is used to draw lines at right angles to an edge. It must always be held on the **face side** or **face edge** (Fig. 3) of the wood with the stock always in the same direction to ensure that all the lines meet. It is also used to check right angles when work is being assembled.

Angles other than right angles are set out or tested with the **sliding bevel** (Fig. 4). The **mitre square** (Fig. 5) is specifically designed for work at 45°. Both are used with the try-square.

The **multipurpose try and mitre square** (Fig. 6) can be used to mark and test both 90° and 45° angles. It is an all metal tool with a sliding head which can be locked into any position along the rule. Its working surfaces are machined to fine tolerances making it an extremely accurate tool.

Fig. 1

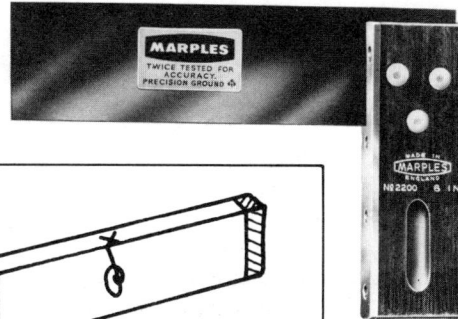

Fig. 2   Try-square

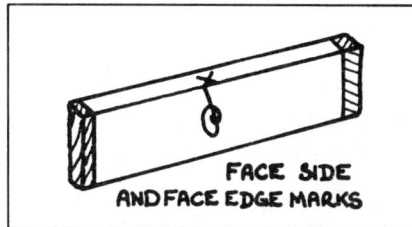

FACE SIDE AND FACE EDGE MARKS

Fig. 3

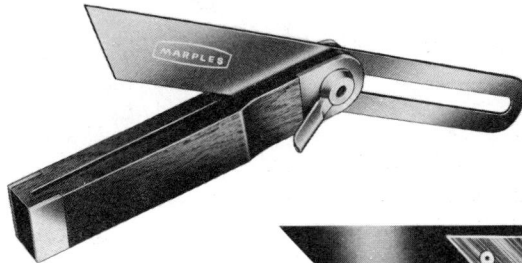

Fig. 4   Sliding bevel

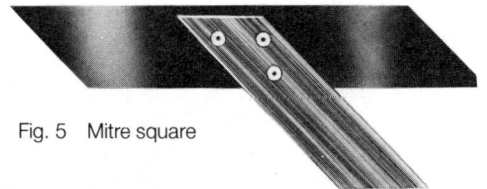

Fig. 5   Mitre square

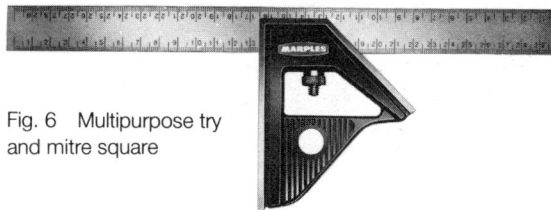

Fig. 6   Multipurpose try and mitre square

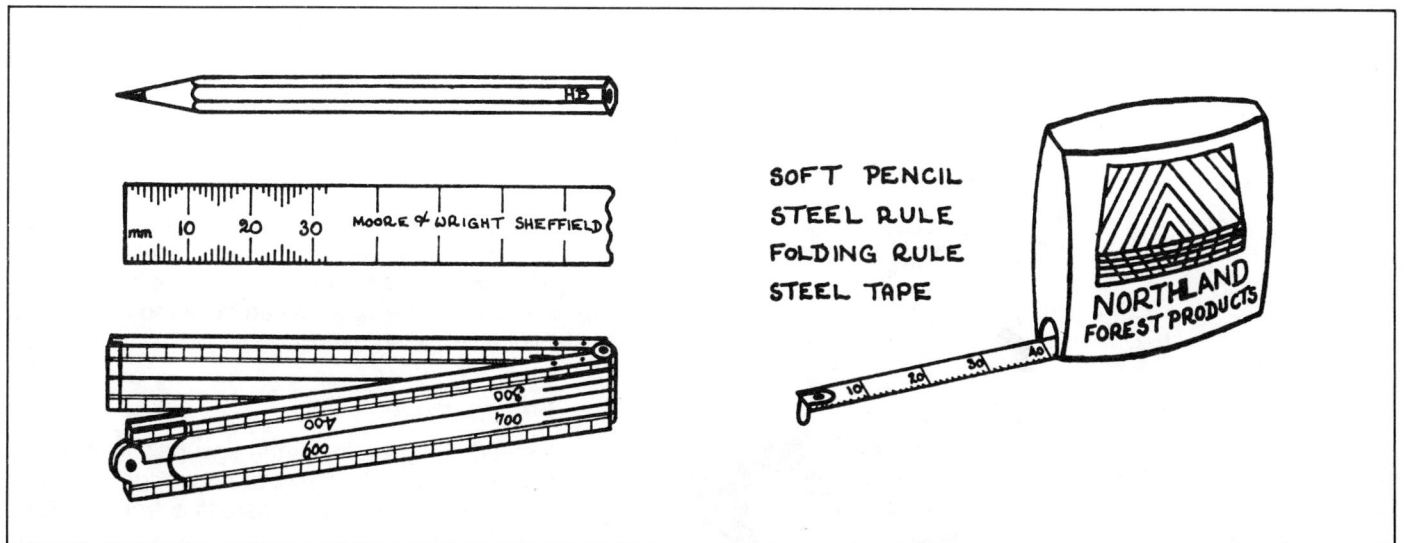

SOFT PENCIL
STEEL RULE
FOLDING RULE
STEEL TAPE

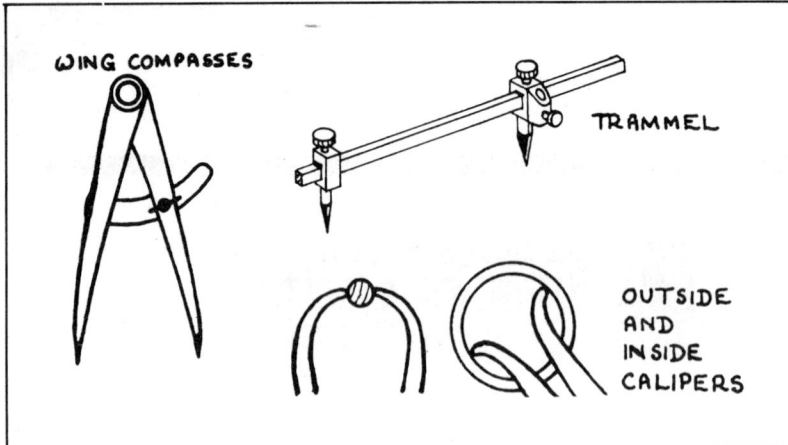

Fig. 7

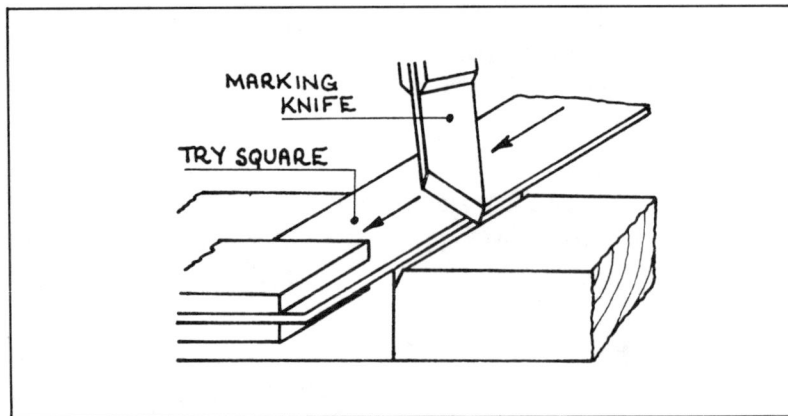

Fig. 8

Wing compasses scribe circles on wood (Fig. 7). Larger circles are drawn with the **trammel**, one head of which has a pencil socket. **Outside and inside calipers** are often used but are particularly necessary for measuring diameters when wood turning.

Lines across the grain of the wood, on which saw cuts have to be made, are often scored with the **marking knife** (Fig. 8). This severs the fibres and the saw produces a cleaner cut. Sawing will also be more accurate because the saw starts easily in the cut which it can then follow. The knife cut must be made along a suitable straight edge, usually the try square.

QUESTIONS

1. List the tools (including pencil and gauges) used for setting out on wood, and rank them, from those used most often to those used least often.

2. Describe the workshop uses of a 300 mm steel rule.

3. (a) Draw a try square and say what its uses are. (b) Why is it such an important tool? (c) How would you test it for accuracy?

4. Sketch and name the tools used for drawing circles on wood.

5. (a) Why is it advisable to make a cut line across a piece of wood before sawing it? (b) Name and sketch the tool with which the cut would be made.

1.2 GAUGES

Woodworkers' gauges scribe or cut lines parallel to the prepared edge or ends of timber. They are used when planing to size, setting out joints, or positioning metal or plastic fittings.

Fig. 9   Single marking gauge

Single marking gauge

The single marking gauge (Fig. 9) is used to scribe lines with the grain of the wood. Its main use is to gauge wood to width and thickness when planing to size, when it should always be held against the face side or face edge. It is made from beech with a boxwood or plastic thumbscrew and steel spur.

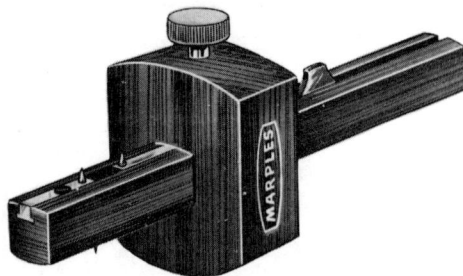

Fig. 10   Mortise gauge

Mortise gauge

The mortise gauge (Fig. 10) has a double spur with which to mark out mortise and tenon joints. The outer spur is fixed but the inner one is on a brass slide, so the two points can be set to the width of the chisel being used.

Fig. 11   Cutting gauge

Cutting gauge

Instead of a spur, the cutting gauge (Fig. 11) has a reversible blade held by a brass wedge. The blade cuts across the grain whereas a spur makes a rough scratch.

QUESTIONS

1. **(a)** Draw a single marking gauge and say what its main use is. **(b)** List three situations when you would use it.

2. What are the differences between the single marking gauge and the mortise gauge?

3. Describe carefully how gauges are held in use.

4. **(a)** Draw simple sketches to show two uses of the cutting gauge. **(b)** Draw a separate sketch to show the blade and wedge assembly.

1.3 SAWS

There are different types of saw, each designed for a particular aspect of woodworking. Because of the grain of the wood, which is formed by its fibrous structure, saws must be capable of cutting **'with the grain'** or **'across the grain'**. Some are sharpened to do one or the other, some do both quite well. The efficiency of all saws depends upon the correct shape and sharpness of their teeth.

Handsaws

Saws for heavy-duty work on large sections of timber are called handsaws (Fig. 12). The blades are wide to maintain the accuracy of the cut, are taper ground from edge to back, and are tensioned so that the blade recovers fully if it flexes in use.

The teeth of all saws are **'set'** alternately to left and right so that the **'kerf'** produced by them is slightly wider than the thickness of the blade and so prevents binding of the blade in the cut.

Handsaws for cutting with the grain are known as **rip-saws** (Fig. 13). Because of the way they are sharpened each tooth acts like a plane or chisel and cuts along the grain of the wood.

Handsaws used to cut across the grain are called **cross-cut saws** (Fig. 14). Because of their 'set' they cut two knife lines across the grain, and as these lines are scored by the movement of the saw, the waste wood crumbles between them. These fragments of wood collect in the gullets between the teeth and fall away as sawdust.

Lines to be sawn 'with the grain' should be gauged; those 'across the grain' marked with a pencil and straight edge i.e. the try square if at right angles, the mitre square if at 45°, or the sliding bevel if inclined at some other angle. These should then be cut with the marking knife before they are sawn.

Tenon saw

The tenon saw (Fig. 15) is one of the cross-cut group, sometimes called backsaws. One purpose of the back is to maintain the rigidity and truth of the deep blade which has been carefully ground and tensioned by the manufacturer. The back also gives it enough additional weight to cut

Fig. 12 Handsaw

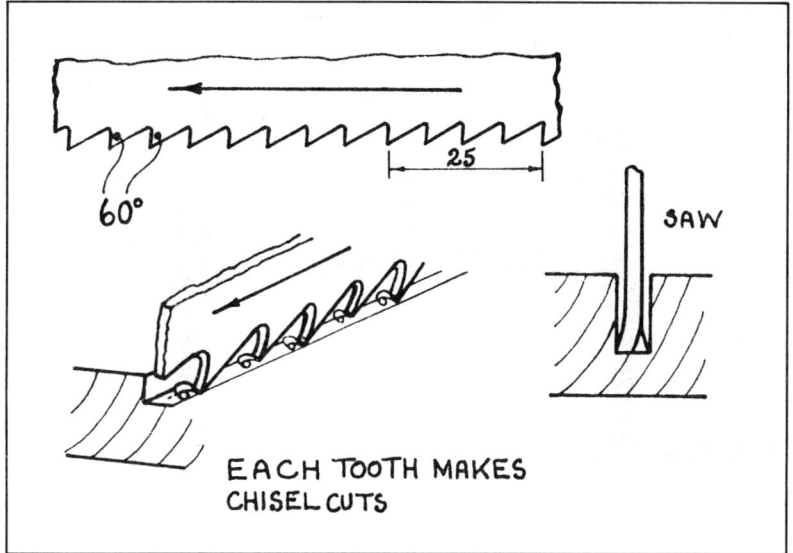

Fig. 13 Rip-saw teeth

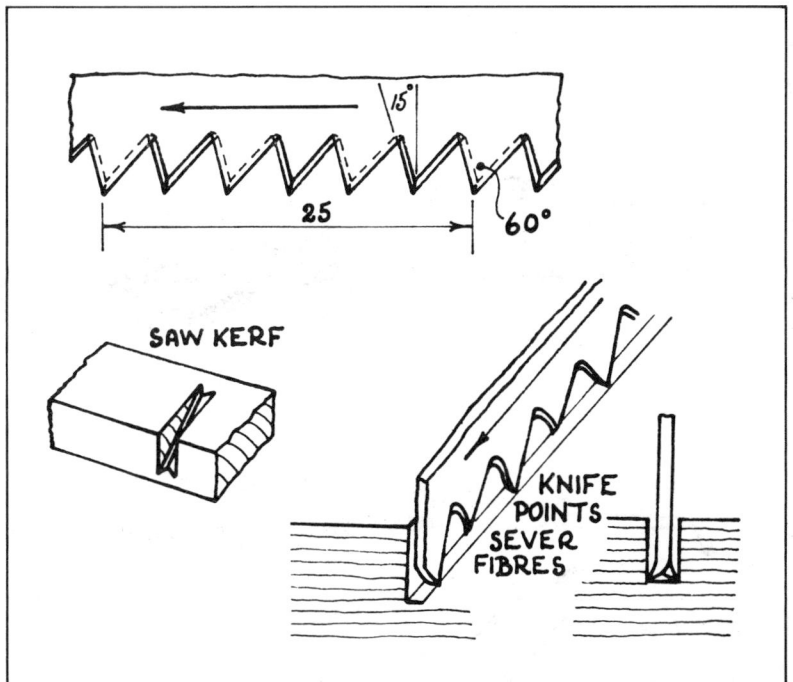

Fig. 14 Cross-cut saw teeth

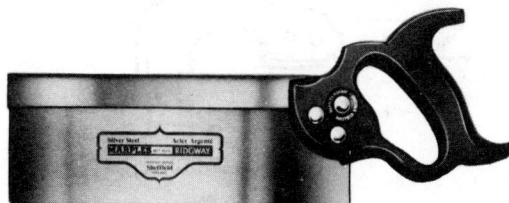

Fig. 15 Tenon saw

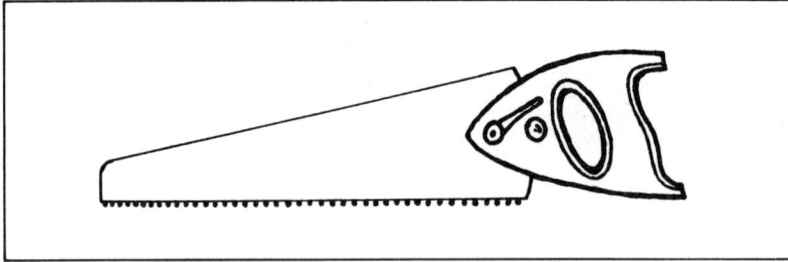

Fig. 16   General purpose saw

Fig. 17   Bow-saw

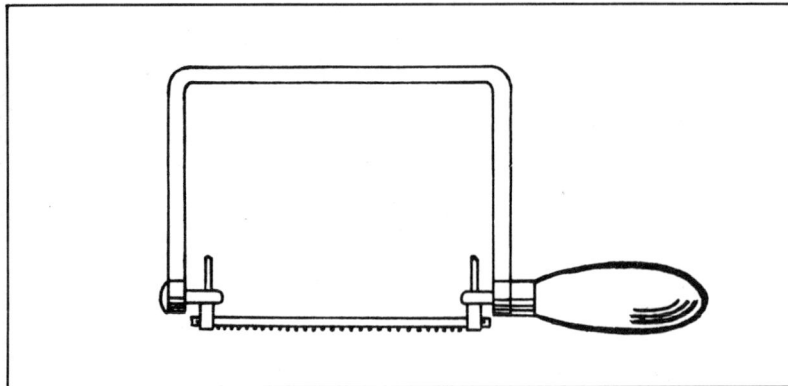

Fig. 18   Coping saw

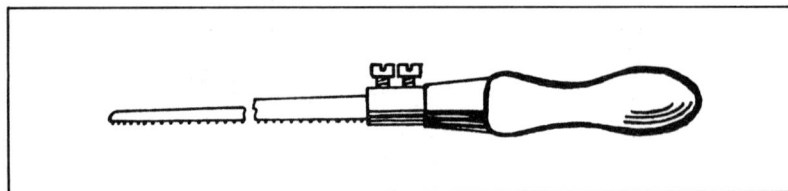

Fig. 19   Pad saw

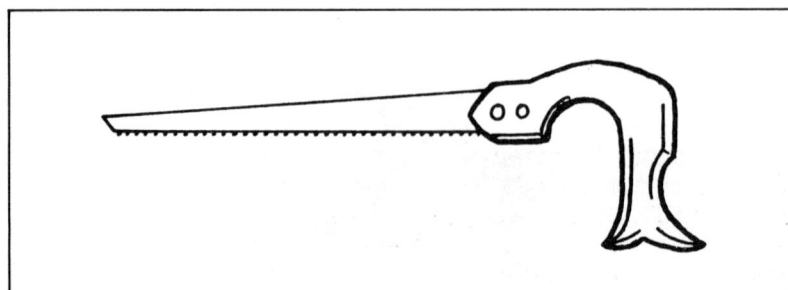

Fig. 20   Compass saw

smoothly through the wood without great pressure having to be exerted. The tenon saw is used mainly at the bench for general frame and carcase construction.

Dovetail saw

The dovetail saw is a small tenon saw, with small teeth and a shorter and thinner blade for cutting small dovetail joints; it is useful for all small work where accuracy is absolutely essential.

When sawing out joints it is always good practice to make the 'long' cuts first, i.e. cuts with the grain, before making the cross-cuts. The waste can then fall away freely as the cross-cut is made, without over-sawing across the grain, so weakening the joint.

General purpose saws

General purpose saws (Fig. 16) have a tough blade with hardened teeth. As well as wood, plywood, chipboard, blockboard, plastics and laminates, they will successfully cut non-ferrous metals and mild steel. The handle can be changed into five different working positions with the quick-release lever.

Bow-saw

The bow-saw (Fig. 17) is used for sawing along a curved profile in quite heavy sections of timber. The handles, and therefore the blade can be rotated through 360°, allowing the frame to clear the workpiece or bench when sawing deep or complicated curved shapes.

Coping saw

The coping saw (Fig. 18) is used for lighter work than the bow-saw. The finer blade will negotiate smaller, sharper curves and more complex profiles with little damage to the workpiece. It can be rotated through 360° within the frame to any required cutting angle. Tension on the blade is maintained by tightening the handle.

Internal cuts can be made by threading the blade through a pre-drilled hole, and then refitting it to the frame. The coping saw is often useful for taking out waste wood when cutting joints e.g. between dovetail pins after the vertical cuts have been made with the tenon saw.

Pad saw

Pad saws or keyhole saws (Fig. 19) are used to cut small internal holes in medium to thick timber. Not having a frame enables the saw to be used at a distance from the edge of the wood. The blades are made from alloy steel and are quite strong. Only the teeth are hardened so the blade remains flexible.

The **compass saw** (Fig. 20) is used in a similar

way to the pad saw but for cutting curves in quite large work.

QUESTIONS

1. (a) Why is a handsaw deep bladed and why doesn't it have a back? (b) Why is the blade thinned towards the back edge?

2. Give reasons for the difference between the teeth of cross-cut saws and rip-saws. Illustrate with simple sketches, showing the direction in which the teeth point.

3. (a) Sketch a tenon saw. What type of saw is it classified as? (b) What is the purpose of the back?

4. When sawing waste wood from a joint why should you make the long cut first? Make simple sketches to illustrate your answer.

5. (a) Name the saws designed for cutting curves in wood. (b) What is their main feature and what is its purpose?

1.4 ABRASIVE TOOLS

Files, rasps and modern Surform tools are in this category. Standard metalworkers' files are used in woodwork, but only to produce a smooth finish, particularly on end grain. Rasps and the Surform are used to rough wood quickly to shape, the bulk of the waste often having been removed before by sawing.

Files

Filing should *always* be with the grain to produce the smoothest possible surface. It is bad practice to file straight across the wood because the far edge can be splintered. However this can be normally avoided by working diagonally. The flat side of the tool is used to produce convex curves, and the curved side to form concave shapes.

Rasps

The **wood rasp** (Fig. 21) is of standard half-round section. Both sides have rasp-cut teeth and the edges are single-cut, which allow it to work into corners. These coarse teeth remove wood very quickly.

Cabinet rasps are thinner and less curved in section than the wood rasp. The teeth are not so coarse so it is suitable for finer work, but it is not edge cut.

Surform tools

The blades are a combination of file and plane (Fig. 22). Each tooth has a cutting edge which chisels or planes away the waste wood. This is cleared immediately through the throat behind each tooth, so the tool does not become choked.

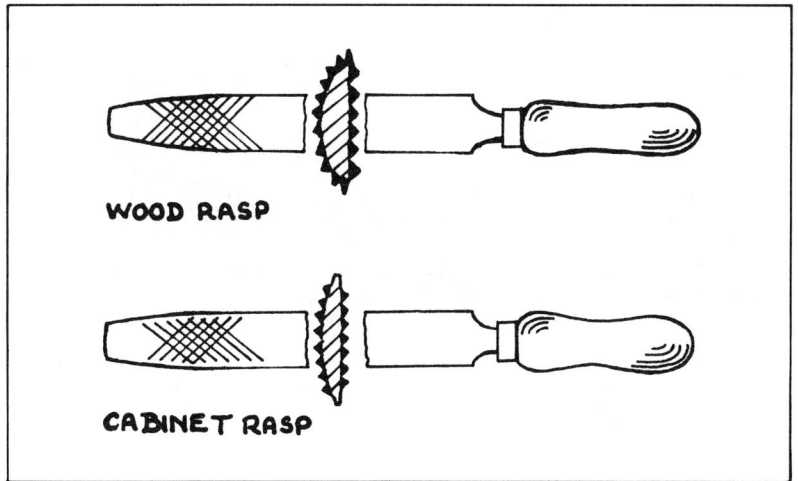

Fig. 21

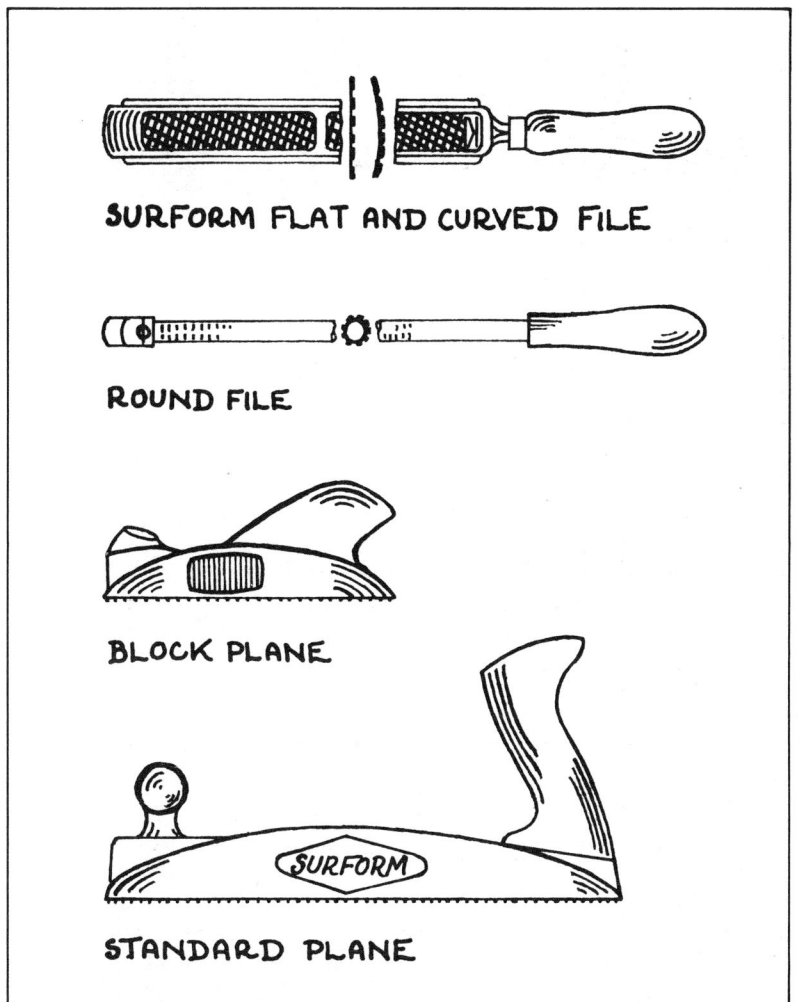

Fig. 22 Surform tools

The blade should always be in tension, but where tension screws are provided they should not be over-tightened. For fast cutting, Surforms should be used at approximately 30° to the direction of cut. They should be used parallel to the cut, for smoothing and fine finishing. These tools are effective on plastic laminates, fibre glass and soft metals, as well as all wood products.

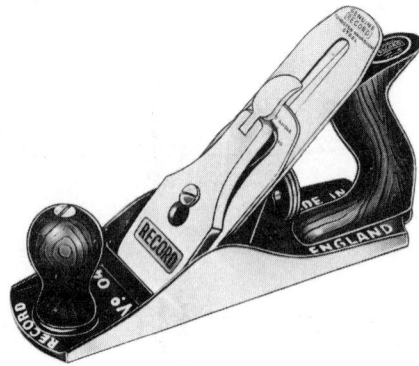

Fig. 23   Smoothing plane

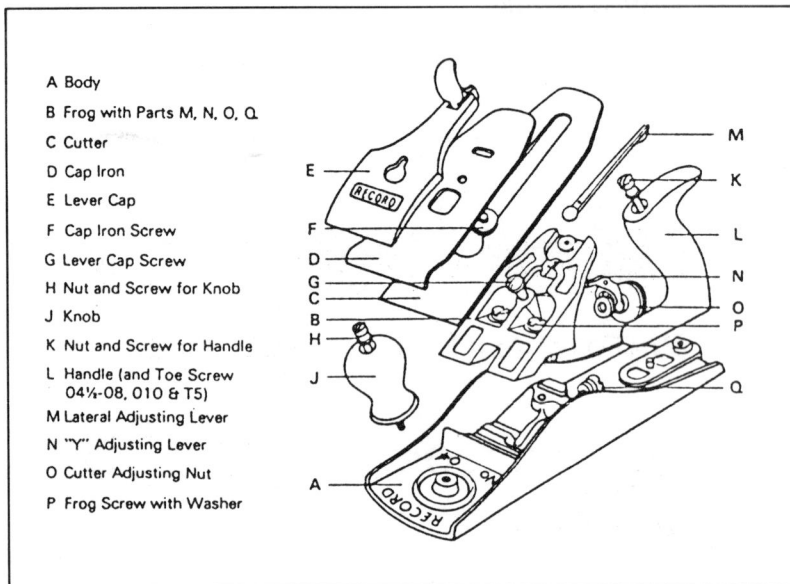

A Body

B Frog with Parts M, N, O, Q

C Cutter

D Cap Iron

E Lever Cap

F Cap Iron Screw

G Lever Cap Screw

H Nut and Screw for Knob

J Knob

K Nut and Screw for Handle

L Handle (and Toe Screw
   04¼-08, 010 & T5)

M Lateral Adjusting Lever

N "Y" Adjusting Lever

O Cutter Adjusting Nut

P Frog Screw with Washer

Fig. 24   Smoothing plane construction

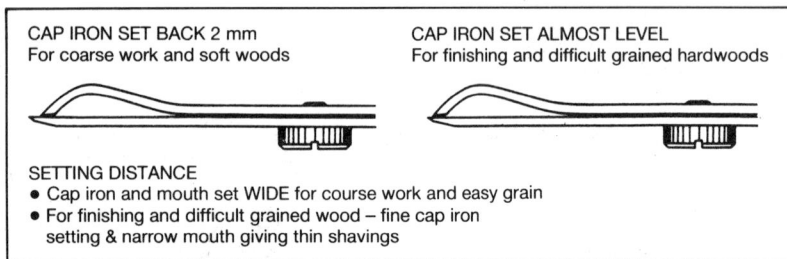

CAP IRON SET BACK 2 mm
For coarse work and soft woods

CAP IRON SET ALMOST LEVEL
For finishing and difficult grained hardwoods

SETTING DISTANCE
• Cap iron and mouth set WIDE for course work and easy grain
• For finishing and difficult grained wood – fine cap iron
  setting & narrow mouth giving thin shavings

Fig. 25   Blade and cap iron

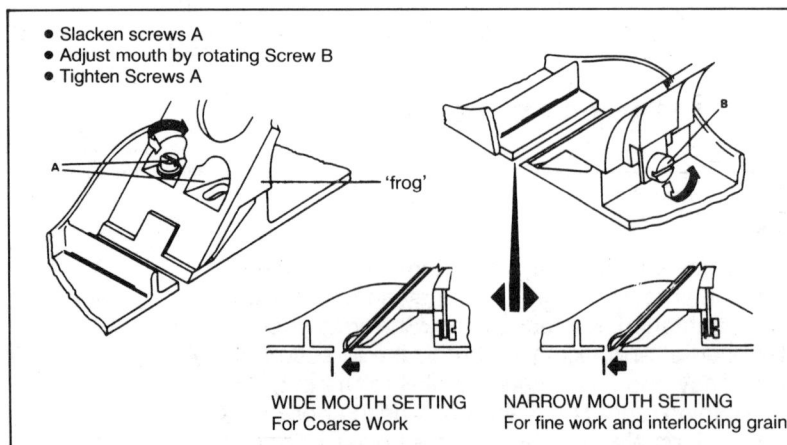

• Slacken screws A
• Adjust mouth by rotating Screw B
• Tighten Screws A

'frog'

WIDE MOUTH SETTING
For Coarse Work

NARROW MOUTH SETTING
For fine work and interlocking grain

Fig. 26   Adjusting the 'frog'

QUESTIONS

1. **(a)** What do you understand by the term abrasive tools? **(b)** Name those which you know and draw one of them.

2. Using sketches and notes illustrate the difference between the wood rasp and the cabinet rasp.

3. Draw an enlarged portion of a Surform blade showing the formation of at least three teeth.

4. **(a)** Sketch three Surform tools. **(b)** Give one example of the use of each.

1.5 BENCH PLANES

The two main purposes of all planes are to reduce the wood to a given size and in doing so produce a smooth surface. Modern planes are of all-metal construction replacing the previous range of planes made from beechwood. Plane bodies are of high grade cast iron. This is a very stable metal which does not twist and wears evenly, so maintaining the truth of the sole.

Smoothing plane

This is the best known of the all-metal bench planes (Fig. 23). Although it is primarily intended for the cleaning up of component parts before assembling and the final finishing of work before sanding, painting, staining and/or polishing, it is also popular for general purpose work. The blade is sharpened so that it removes wide thin shavings from the surface of the wood.

The construction of the smoothing plane (Fig. 24) and the mechanics by which the blade is adjusted are common to all the planes of this type.

The blade and its cap iron (Fig. 25) are held together with a large cheese-head screw. The purpose of the cap iron is to lift the shaving and guide it out of the plane. It is important that the cap iron fits tightly across the blade, so that shavings do not become wedged between the two and choke the mouth. It is equally important to set the cap iron a suitable distance back from the cutting edge to get the best possible finish on the wood being planed, depending upon its particular texture and grain.

The blade assembly is clamped to the 'frog' (Fig. 26) by means of the lever cap iron and for the smoothest possible finish the width of the plane mouth should be varied, again depending upon the nature of the wood. This setting is made by adjusting the 'frog'.

The final setting of the blade (Fig. 27) is made with the **lateral adjusting lever** (1) and **cutter adjusting nut** (2). The lever is used to bring the cutting edge parallel with the sole, and the adjusting nut is used to regulate the depth of cut, i.e.

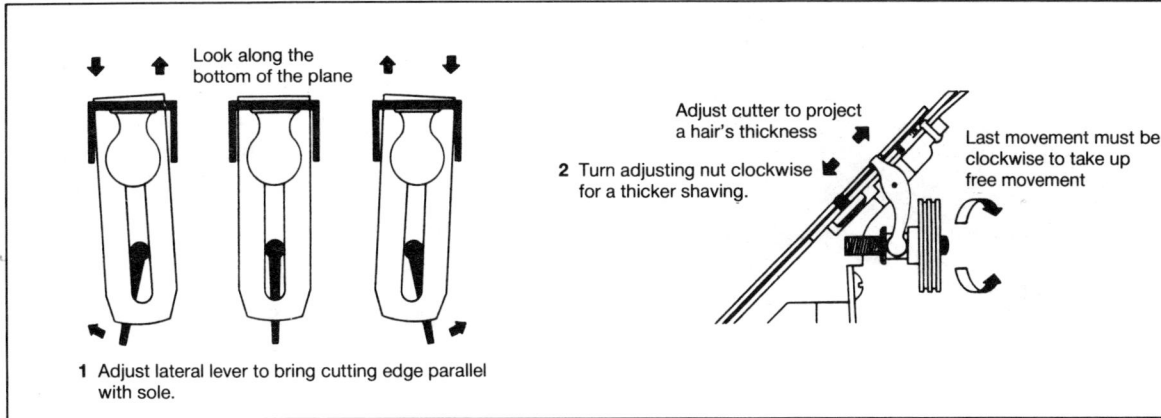

Fig. 27 Setting the blade

thickness of the shavings. These adjustments are checked by looking along the sole of the plane.

Jack plane and jointer plane

The other planes in the group are the jack plane and the fore or jointer plane (Fig. 28). They are considerably longer than the smoothing plane.

The jack plane is used for the preliminary squaring of rough timber. It is the plane generally chosen for the quick removal of waste wood, as the blade is sharpened to a slight curve across its width. Unfortunately because of this, curve ridges are left on the wood which are very unsightly and must be removed with the smoothing plane.

The fore or jointer plane is used to produce very accurate surfaces on long timber e.g. when fitting doors.

For the best results it is important to select the correct plane for each stage of working (Fig. 29).

Metal planes tend to 'drag' on the wood rather more than wooden ones. This is more apparent as the wood becomes smoother and flatter. It is particularly noticeable on resinous woods and to overcome this, most of the above planes are available with **corrugated soles** which reduce the friction between the plane and the wood (Fig. 30).

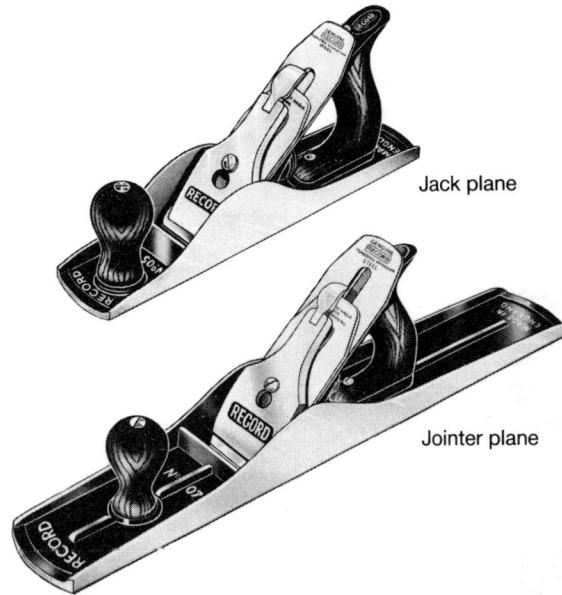

Fig. 28

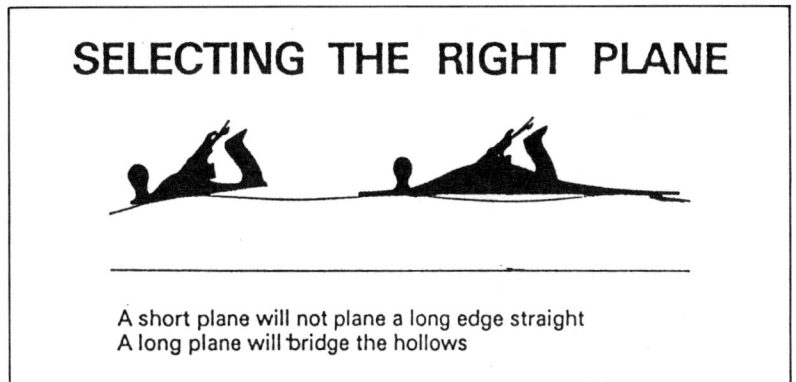

Fig. 29

QUESTIONS

1. (a) Why is cast iron used in the construction of modern planes? **(b)** Because of the nature of cast iron, are there any precautions you should take when using these planes?

2. (a) Which plane would you use to plane the wood to size for a small stool? **(b)** Give the reasons for your choice.

3. How would you set a smoothing plane to clean up the sides of a small box before gluing up? Use simple sketches to illustrate your answer.

Fig. 30 Corrugated sole

Fig. 31   Rebate plane

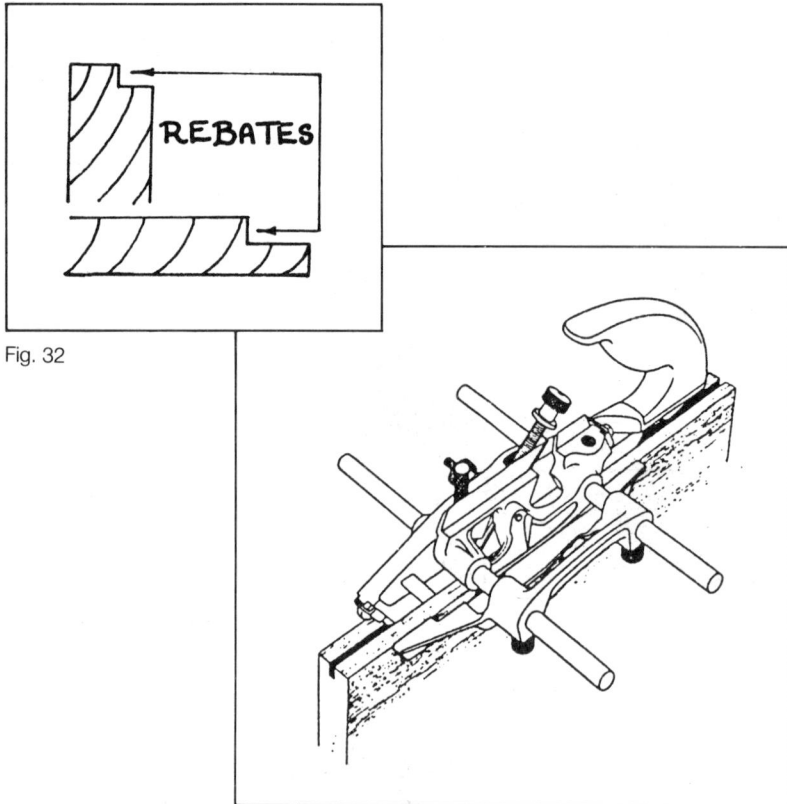

Fig. 32

Fig. 33   Plough plane cutting a groove

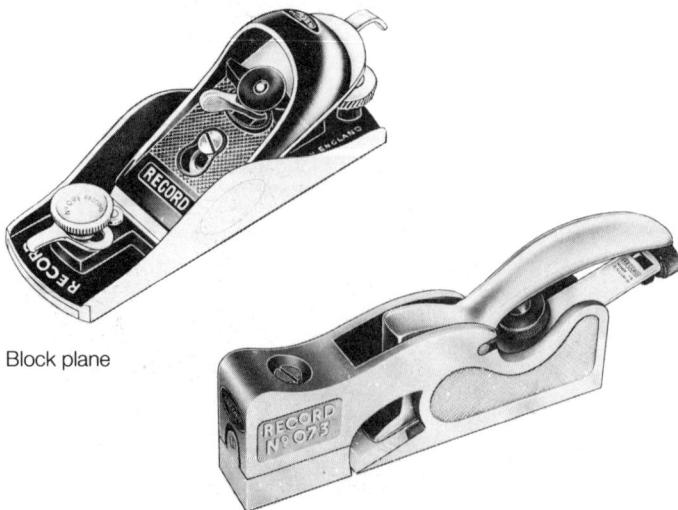

Fig. 34   Block plane

Fig. 35   Shoulder rebate plane

4. (**a**) What is the purpose of the cap iron on the bench plane? (**b**) In what way can it be adjusted? Why is this necessary?

5. The cutting edge of the jack plane blade should be sharpened to a different shape to that of the smoothing plane. Illustrate the difference and say why this is done.

1.6 SPECIAL PURPOSE PLANES

Sometimes wood needs to be shaped to complex shapes for which a certain number of special purpose planes are available.

Rebate plane

The rebate plane (Fig. 31) is used to work out 'square cuts' or 'laps' (rebates) at the edge or end of a piece of wood (Fig. 32). The waste wood should be marked off carefully, preferably with a single marking gauge. The adjustable depth stop and adjustable fence are then set to these lines. There is a spur to sever the fibres ahead of the blade, which is particularly necessary if working 'across the grain'. The adjustable fence can be used on either side of the plane which simplifies setting, and accurately controls the size of the rebate. There are two cutter positions, the standard one for normal rebating and a forward one for planing into corners and taking out 'stopped rebates'.

Plough plane

The basic plough plane (Fig. 33) is used for cutting grooves as its name suggests. Similar in many ways to the rebate plane, a simple conversion kit of cutters gives it increased working capacity to include rebating, tongue and grooving, fluting, beading and reeding.

Block plane

Because block planes (Fig. 34) are short they are convenient for one-handed use, and are ideal for use on small fine workpieces where accuracy is absolutely essential. They are also very effective for small scale chamfering and bevelling, and because of the low angle at which the blade is set are particularly efficient for planing end grain.

Shoulder rebate plane

Sometimes known as the cabinet makers' plane (Fig. 35) it is designed for very fine, accurate work, and is small and compact for one-handed use. The base and sides of the body are precision machined so the plane can be used when laid flat on either side, as well as for normal planing. Because the blade is set very low and the plane is heavy, being made from cast iron, it is ideal for planing end grain. As its name suggests, it is used across the shoulders of larger tenons which may need to be trimmed.

Bull-nose shoulder plane

This is similar to the shoulder plane but less than half its length (Fig. 36). It is intended for very fine, accurate work and can be used as a rebate or shoulder plane. By removing the nose it is converted into a chisel plane with which to complete stopped rebates and chamfers (Fig. 37).

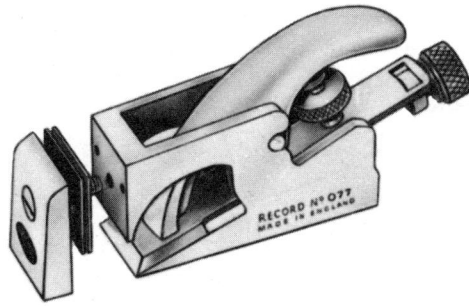

Circular plane

Sometimes called a compass plane (Fig. 38), it is designed for use on concave and convex surfaces, the curve of the flexible, steel sole being set accurately by the centre screw adjustment. Whenever possible planing should be with the grain. The curved frame provides a comfortable grip at each end.

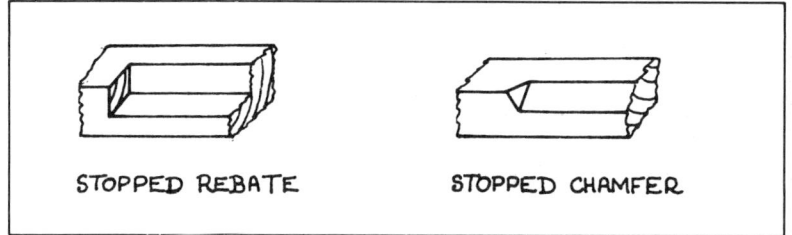

Spokeshaves

These tools were originally designed to shape the spokes of wooden wheels and were made of wood. The modern spokeshave (Fig. 39) is made from a malleable iron casting and has a tool steel blade with adjusting screws for depth of cut. Round soled and flat soled spokeshaves are made to produce concave and convex curves respectively. The tool should always be held firmly, and normally at right angles to the work. However, there are occasions when a smoother cut and better finish can be obtained if it is slightly angled, care being taken to avoid cutting against the grain.

Fig. 36 Bull-nose shoulder plane

STOPPED REBATE   STOPPED CHAMFER

Fig. 37

Fig. 38 Circular plane

Fig. 39 Spokeshave

QUESTIONS

1. You have several long boards from which to make a table top. **(a)** Which plane would you use to true the edges? **(b)** How would you test and mark them?

2. List in order the steps in setting up a rebate plane ready for use.

3. **(a)** What is a plough plane and what is its special function? **(b)** How can this be varied and for what purposes?

4. **(a)** Which plane would you use to true the shoulders of a wide tenon? **(b)** Why is this plane particularly suitable for this purpose?

5. Some special planes have their blades set at a very low angle. What are the advantages of this?

1.7 SCRAPING TOOLS

Router

The purpose of the router (Fig. 40) is to flatten the bottom of 'through' and 'stopped' housings,

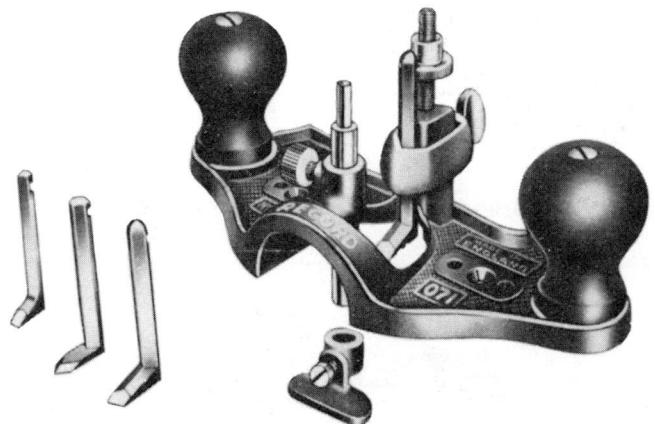

Fig. 40 Router and cutters

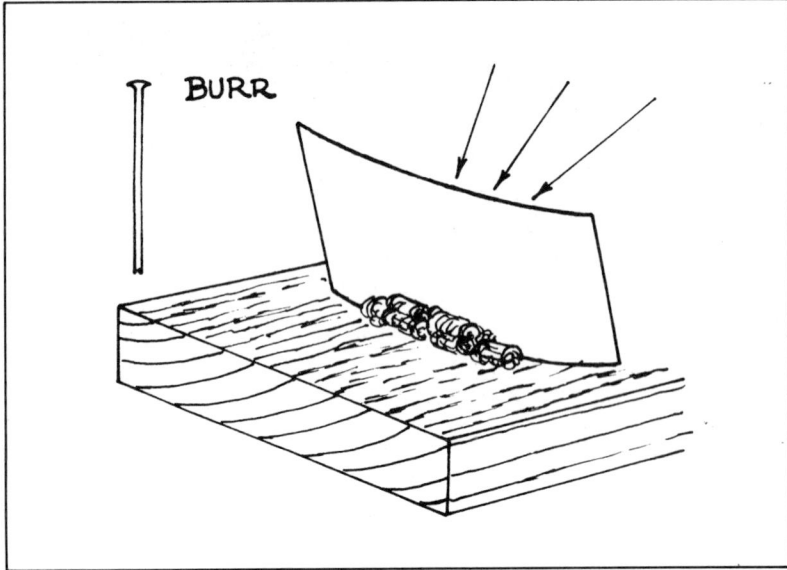

Fig. 41   Cabinet scraper

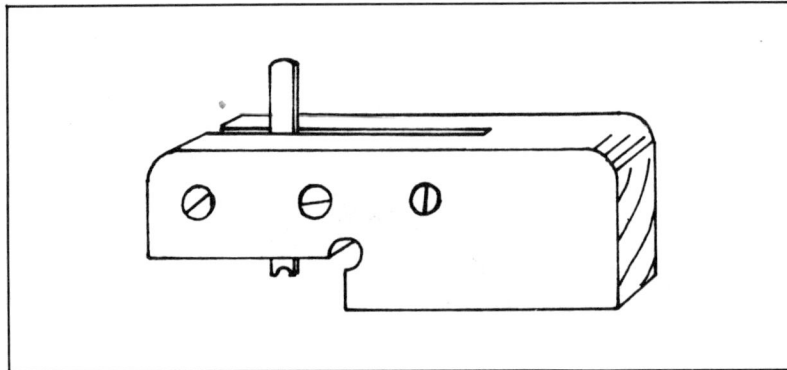

Fig. 42   Scratch stock

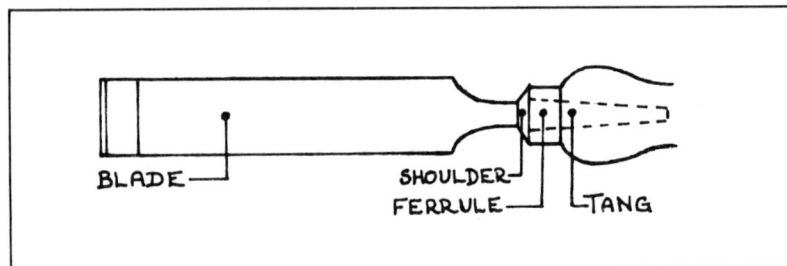

Fig. 43

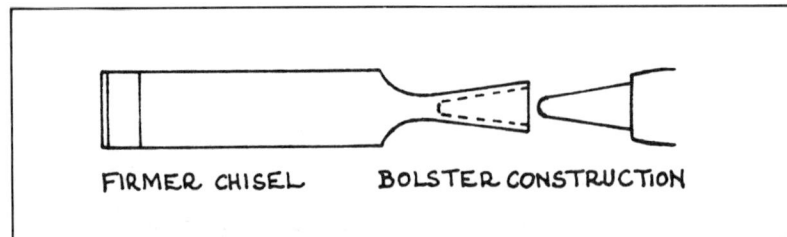

Fig. 44

Fig. 45   Firmer chisel

which it does by scraping rather than cutting, because the blade is always set or sharpened at a very steep angle. The workpiece must be clamped down firmly and the bulk of the waste removed by first sawing and then chiselling. The router blade is then put in the groove and the body held in both hands and tipped forward. By drawing the router forward and backwards the cutter scrapes out the remaining waste. The body of the tool gradually settles down on the board, leaving the housing at the correct depth.

Cabinet scraper

These simple tools (Fig. 41) will provide a smooth finish on the most difficult hardwoods, and improve on the surface produced by the smoothing plane where the grain is interlocked or curly, or where there are knots. Pressure is applied to the centre of the scraper blade with the thumbs. It is then pushed along the wood, slightly tilted forward so the burred edge removes fine shavings. When the irregularities have been scraped off, the workpiece can be glass-papered in the usual way.

Scratch stock

These are home-made tools (Fig. 42), usually made only when the need occurs. Using these, small decorative reeds, flutes and grooves can be scratched out as required.

QUESTIONS

1. Draw a simple router and describe its use.

2. Using sketches and simple notes show how the router blade is set to the correct depth.

3. (a) Design a cabinet scraper suitable for smoothing the inside of a shallow elliptical dish. (b) Draw an enlarged section through it to show the burred cutting edges.

1.8 CHISELS AND GOUGES

Chisel size is measured across the cutting edge of the blade. Blades are made of hardened and tempered tool steel, accurately ground to a fine finish on all sides and edges. **Ferrules** (Fig. 43) are fitted to prevent the handle from splitting. Now these are often made of steel rather than traditional brass. The blade is fitted to the handle by means of a **tang** which penetrates as far as the shoulder. Many modern chisels are however of **bolster construction** (Fig. 44), the handle fitting into a tapered socket on the blade. Handles may be made from ash, boxwood or plastics, and there is a range of shapes from which to choose.

Firmer chisel

The thick blade gives the chisel its name (Fig. 45).

It is used for work where a strong rigid blade is required, and where the danger of undercutting has to be avoided, e.g. when cutting grooves with 90° edges. The chisels are strong enough to be used with the mallet for joint cutting.

Bevel-edge firmer chisel

These chisels (Fig. 46) are intended primarily for fine cabinet work, dovetailing, and working in angled corners where the blade of the firmer chisel would be too thick. However, what was originally a specialist tool is now manufactured with additional strength and can be used as a general purpose chisel.

Paring chisels (Fig. 47) have a similar blade shape, but are thinner and longer to give a good reach, e.g. when cutting long grooves and housings.

Mortise chisel

These chisels are used with the mallet for chopping out joints and slots. The cuts are always across the grain. Therefore the blades have to be very sturdy to withstand the mallet blows and leverage applied in removing the waste wood. The thickness of the blade stops the chisel from twisting in the mortise, maintaining the truth of the cut.

There are three main types of mortise chisel.

Registered pattern chisel
This is a traditional chisel (Fig. 48) with a strong stout blade fitted into a round, ash handle bound at each end with steel ferrules. The top ferrule prevents splintering of the wood from the mallet blows, the lower one, cushioned by a leather washer, takes up the pressure from the tang.

Splitproof mortise chisel
A chisel designed for deep mortising, it has an impact resistant handle which is guaranteed for life. The handle shape makes it comfortable to use and because it is broad, the risk of foul blows is reduced. It has a very thick blade to withstand heavy leverage in deep holes.

General heavy duty chisel
Classed as a 'construction worker's' chisel by the manufacturers (Fig. 49) this has the same impact resistant handle as the Splitproof mortise chisel. The makers guarantee the handle against splits from hammer and mallet blows.

Use of the chisel

Always use a sharp chisel for easy cutting and to get a good finish. A blunt chisel is dangerous, difficult to guide and requires too much pressure to make it cut.

Use a slicing action and cut with the grain. It is better to take off two thin shavings than one thick one.

Use as strong a chisel as is suitable for the

Fig. 46   Bevel-edge firmer chisel

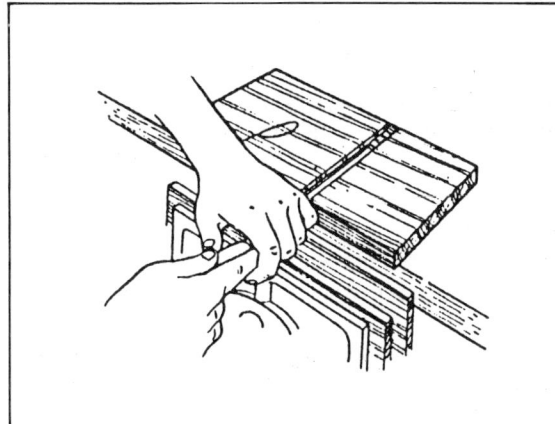

Fig. 47   Paring chisel

Fig. 48   Registered pattern chisel

Fig. 49   General heavy duty chisel

Fig. 50   Out-cannel gouge

Fig. 51   In-cannel gouge

work. Preferably use a mallet when striking the chisel. Hold the work firmly in a vice, or with a clamp or bench hold fast.

In use the chisel should always be held vertically unless a special angled cut is required.

Firmer gouges

Gouges are used in the same way as chisels but because of their rounded section, cut curves instead of flat surfaces. When the gouge is ground and sharpened on the outside it is referred to as **'out-cannel'** (Fig. 50). With its scooping action it can be used to hollow out recesses, wooden bowls and dishes, and many forms of wood sculpture. **'In-cannel'** gouges (Fig. 51) are ground on the inside of the blade, and are often called scribing or paring gouges. Their uses include cutting mouldings, and paring out short grooves and flutes either 'with the grain' or 'across the grain'.

QUESTIONS

1. **(a)** How is chisel size specified? **(b)** Draw cross-sectional sketches to show the blade shape of three different types of chisels and give their names.

2. When would you use a bevel-edge chisel in preference to a firmer chisel? Sketch two examples.

3. List examples of the use of the firmer chisel.

4. Sketch a bevel-edge chisel and name the various parts.

5. The mortise chisel has to be particularly strong. **(a)** Explain why this is so. **(b)** Make sketches to show why it is strong.

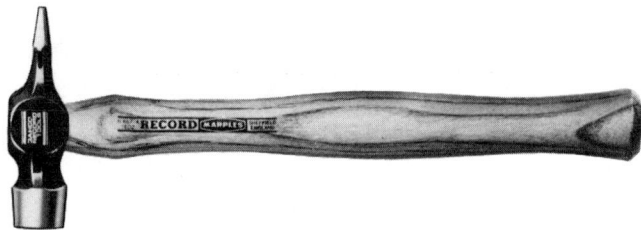

Fig. 52   Joiner's or Warrington pattern hammer

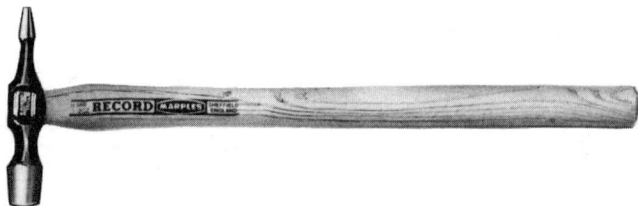

Fig. 53   Pin or tacking hammer

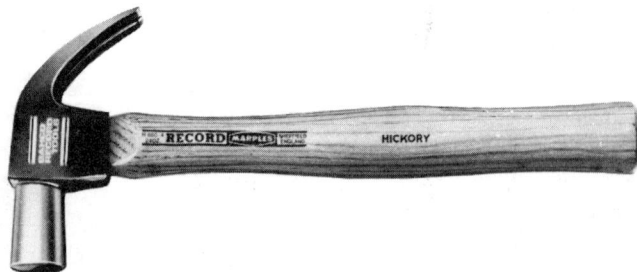

Fig. 54   Claw hammer

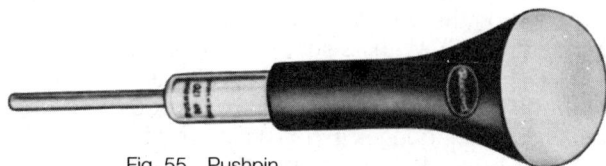

Fig. 55   Pushpin

6. **(a)** Distinguish between 'in-cannel' and 'out-cannel' gouges. **(b)** Draw simple diagrams of their cutting edges.

7. Describe two uses of each of the above gouges and illustrate with sketches.

1.9 HAMMERS AND MALLETS

Hammers

Hammer-heads are drop-forged from high quality carbon steel. American hickory is the best wood for hammer shafts as it is more resilient and will withstand heavy use even better than ash. A loose hammer-head or one which detaches completely is a safety hazard which can cause serious injury. Manufacturers make every effort to avoid this, usually by using a system of wooden and metal wedges, to fix the head firmly to the shaft.

Joiner's or Warrington pattern hammer

This is a cross-pein hammer (Fig. 52); the narrow edge is set at right angles to the handle but slightly off-centre to the axis of the head. It can be used to start small nails, or for work in restricted spaces such as grooves and corners. The round pein has a good surface area which registers well on the nail head.

Pin or tacking hammer

This is a small hammer (Fig. 53) for use on very light workpieces where the slim head is an advantage. It does not obscure the work and small nails can be driven home in a restricted space. The handle is made of ash and is almost straight, to retain as much strength as possible.

Claw hammer

Designed for heavy working (Fig. 54), the hammer gets its name from its extractor claws. The shaft is made of hickory to withstand the leverage applied to it. The round pein should be slightly domed so that the force of the blow is concentrated on the nail head. This reduces the possibility of hammer marks being left on the wood.

Pushpin

Panel pins up to 32 mm long can be driven through hardboard or plywood quickly without bending them, with this modern tool (Fig. 55). It is not recommended for thick plywood or for pinning into hardwoods. The pin is loaded into the brass barrel where it is held by a magnetic plunger. A firm push drives the pin home and the spring action leaves the tool ready for reloading. It is particularly useful where there is limited space.

Mallets

These are simple, well designed tools. The heads are usually made from beechwood, and the handles from beech or ash. The beech head delivers a sound blow to the chisel, and the handle is sufficiently resilient to absorb the shock. Care should be taken to avoid excessive bruising of the head or this may result in foul blows to the chisel.

Wood carver's mallet

The round beechwood head (Fig. 56) provides a constant striking surface, and although round in section it can be used with great accuracy.

Joiner's or cabinet maker's mallet

This mallet (Fig. 57) is heavy enough to drive chisels into the wood when cutting joints but can deliver light blows when needed. The striking faces are inclined so the blow is delivered squarely to the chisel, and the shaft and socket are tapered so the head becomes more firmly wedged when in use. It is also very useful when work is being assembled, the broad face being less likely to make marks on the wood than a hammer.

QUESTIONS

1. Draw simple sketches to show the difference between the heads of Warrington, pin, and claw hammers.

2. (a) From which woods are hammer shafts made? (b) Why are these woods used and which would you prefer to use?

3. When buying a hammer you should examine the shaft closely. What specific details should you look for?

4. Draw a cabinet maker's mallet. (a) How is the head fixed? (b) From what is the mallet made?

5. How does a wood carver's mallet vary from the above? Illustrate your answer.

1.10 SCREWDRIVERS

There are now many tools of this type. Apart from screwdrivers other similar tools are used for the different screw recesses now available. All these tools rely on torque applied with the hand or hands for their effective use. Many screwdrivers now have plastic handles.

Cabinet screwdriver

This is a traditional screwdriver (Fig. 58) still preferred by many craftsmen. The oval section handles of polished boxwood with their bulbous ends are designed for maximum comfort and

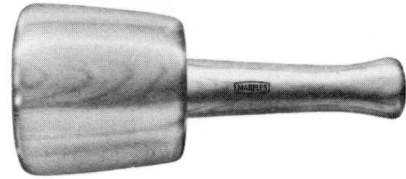
Fig. 56 Wood carver's mallet

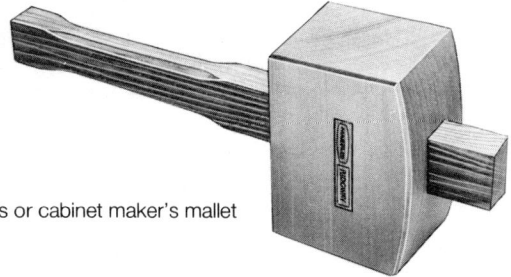
Fig. 57 Joiner's or cabinet maker's mallet

Fig. 58 Cabinet screwdriver

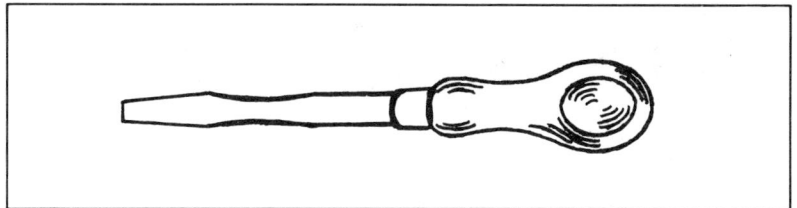
Fig. 59 London pattern screwdriver

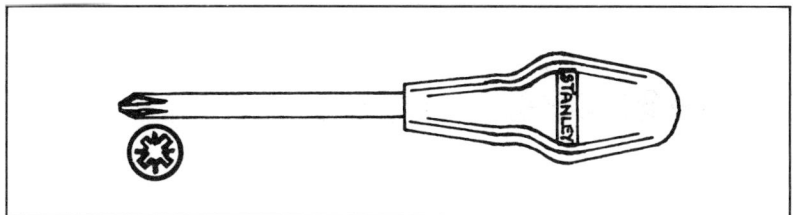
Fig. 60 Pozidriv screwdriver

positive grip, so giving positive torque. Blades are made from forged chrome-vanadium steel, with flared tips specifically designed for use with slotted wood screws.

London pattern screwdriver

This (Fig. 59) has a strong blade forged from thick flat-sectioned steel. It is a very robust tool; the flat-sectioned handle combined with the wide tip of the blade make it very efficient, particularly for turning heavy gauge screws.

Pozidriv screwdriver

This screwdriver (Fig. 60) is used for cross-headed screws. These drivers will also fit the new **Supadriv** recess, the angle of which is slightly different to the Pozidriv, so the screwdriver is less likely to 'ride out' in use.

Fig. 61   Bradawl

Fig. 62   Chisel-pointed bradawl

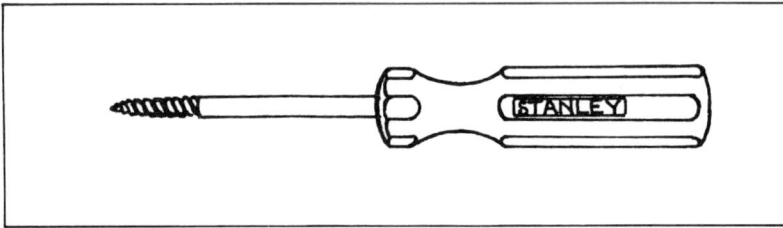

Fig. 63   Screwstart

Fig. 64   Hand drill or wheel brace

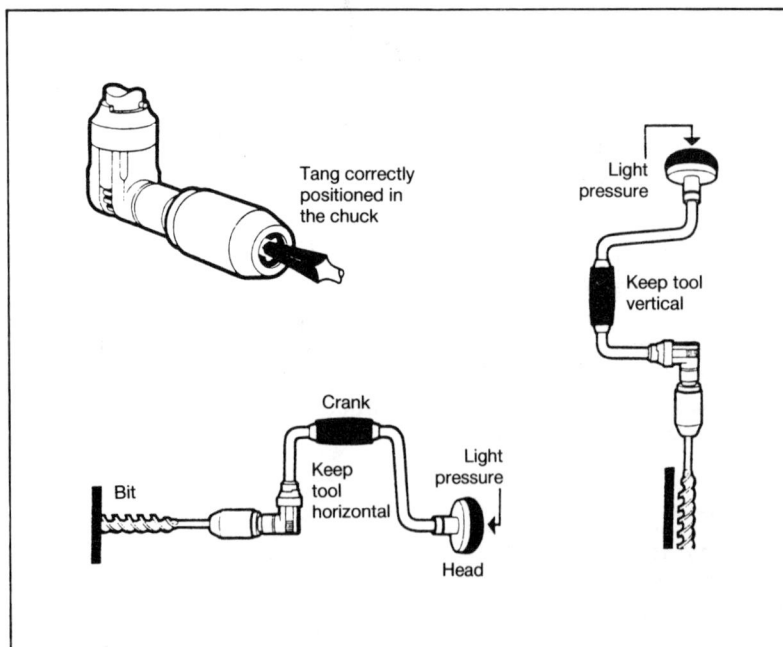

Fig. 65   How to use a joiner's brace and bit

Tang correctly positioned in the chuck

Light pressure

Keep tool vertical

Crank

Keep tool horizontal

Light pressure

Bit

Head

Bradawl

A bradawl (Fig. 61) is used to make holes in which to start screws. To avoid splitting the wood, press the cutting edge 'across the grain'. The bradawl with the square section of the blade drawn down to a fine point, allows quicker and easier withdrawal than the chisel-pointer bradawl (Fig. 62).

Screwstart

A lead thread can be cut in the workpiece with this tool (Fig. 63). This is then 'picked up' by the fixing screw. When metal or plastic fittings are being fixed, they can be held in place and the screw thread started, using a screwstart. If wooden members are being screwed together, the tool can be put through the 'clear' hole in the top piece and the thread started in the second piece below.

QUESTIONS

1. Draw clear sketches to show how the tip of a screwdriver blade should fit into a slotted screwhead.

2. Show how screwdriver blades are fitted (**a**) to wooden handles (**b**) to plastic handles.

3. Draw the cross sectional shape of a screwdriver handle to show how the required torque is applied to the screw.

4. List the advantages of recess-headed screws.

5. (**a**) Draw a chisel-pointed bradawl. (**b**) How does it work?

1.11 WOOD BORING

For successful woodworking it is very important to be able to bore holes accurately. With the **pedestal drilling machine** (See 1.13) complete precision is possible. While the **portable electric drill** (See 1.13) takes much of the effort out of drilling, more care is needed to ensure accuracy. With this drill a number of holes can be drilled quickly, either at the bench or 'on site'. The hand drill or wheel brace is slower but for small holes gives good results. With any of the above, metal-worker's twist drills make good, clean holes in wood.

Hand drill or wheel brace

Originally the hand drill (Fig. 64) was considered to be an engineer's tool for drilling small diameter holes in small items of metalwork. It is now seen as a very convenient tool for doing the same work in wood or plastics. Holes to take nails and screws can be drilled very accurately. Hand drills are

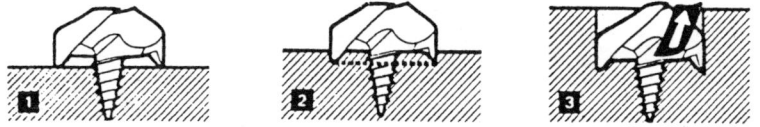

HOW A BIT BORES

1. Screw point pulls the bit into the timber.

2. Spurs scribe the diameter of the hole in advance of the cutters.

3. Cutter lifts the chips which then pass up the twist.

Fig. 66

fitted with a 3-jaw self-centering chuck into which the drill is fitted.

Joiner's brace

Modern drilling tools have largely replaced the traditional joiner's brace (Fig. 65). However it still has a place where large diameter, deep holes are needed and where there is room to swing the crank.

The tang of the bit is held in the jaws of the chuck. They are spring-loaded and controlled by the chuck shell which in turn screws to the end of the crank. The crank is fitted with a centre grip which is free to rotate. Cutting pressure is applied through the head which runs on ball bearings. If brace bits are in good condition, only light pressure should be necessary to bore the hole because the screw point pulls the bit into the timber (Fig. 66).

Double-ended screwdriver bit

This bit (Fig. 67) is made from tough high grade carbon steel, to withstand the strain of turning large heavy gauge screws. The increased torque of the brace makes it possible to drive screws home tightly and quickly.

Original centre bit

The spur scribes the hole (Fig. 68), cutting through the fibres as it rotates round the long central point. The single cutter then lifts out the waste wood. The rate of cut depends entirely on the pressure applied to the brace.

Quick-cutting centre bit

The bit (Fig. 69) is drawn into the wood by the screw point, but otherwise the cutting action is the same as for the original centre bit. As the name suggests this bit cuts more quickly than the original centre bit.

Centre bits are not suitable for drilling deep holes as they tend to run off-centre because they have no length of twist to guide them.

Forstner bit

Unlike other bits (Fig. 70) it is guided by its circular rim and not by the centre point. It will bore any arc

Fig. 67 Double-ended screwdriver bit

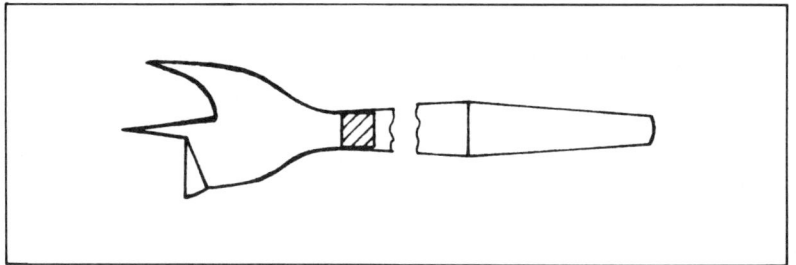

Fig. 68 Original centre bit

Fig. 69 Quick-cutting centre bit

Fig. 70 Forstner bit

of a circle and is unaffected by knots or run of the grain. The holes it bores are clean, true and flat-bottomed and can be overlapped without difficulty.

Fig. 71   Jennings pattern

Fig. 72   Scotch auger

Fig. 73   Solid centre bit

Fig. 74   Rosehead countersink bit

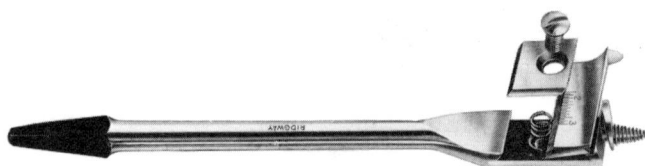

Fig. 75   Expansive bit

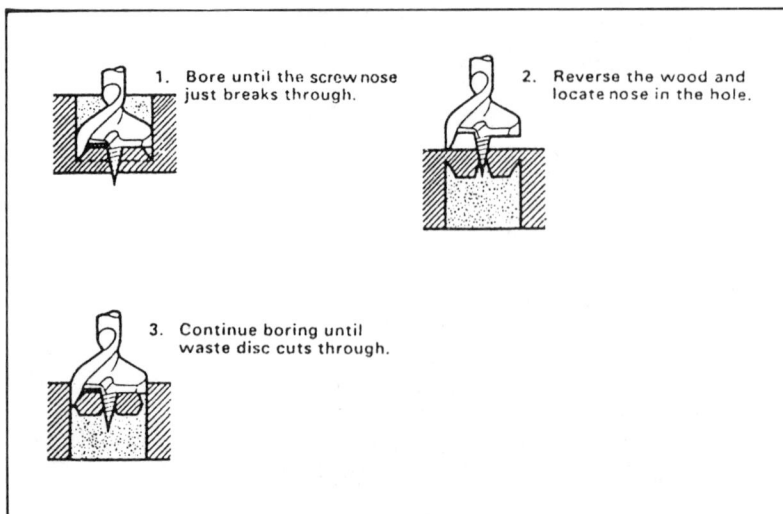

1. Bore until the screw nose just breaks through.

2. Reverse the wood and locate nose in the hole.

3. Continue boring until waste disc cuts through.

Fig. 76   Boring 'through' holes

Auger or twist bit

The long twist guides the direction of the bit as it bears on the side of the hole keeping it accurate. It will therefore bore in any direction through the wood. Three types of auger bit are made:

Jennings pattern perfectly balanced with two spurs, two cutters and a double twist to lift and throw out waste (Fig. 71). In general they are preferred by craftsmen and cabinet-makers because they give such good results.

Scotch auger does not have any spurs and is designed and sharpened for cutting hardwoods (Fig. 72). It is particularly effective for boring down end grain.

Solid centre bit ideal for general use where fine cutting is not important (Fig. 73). It has two spurs and cutters and only one spiral but is very strong because of the solid centre.

Rosehead countersink bit

This is for recessing holes to take the head of a screw so that it is level with or slightly below the surface of the wood (Fig. 74).

Expansive bit

Cuts (Fig. 75) in a similar way to the centre bit. A locking screw and a wedge with a serrated edge are used to adjust the cutter. Like the centre bit it is only effective on thin sections of wood.

Special care must be taken when boring 'through' holes as opposed to 'stopped' ones (Fig. 76). The hole must be bored from each side or into a waste block clamped to the back, to avoid too much splintering.

QUESTIONS

1. (a) Distinguish between the original centre bit and the quick-cutting centre bit. (b) How does each work?

2. (a) Sketch a Jennings pattern auger bit. (b) Why is it so efficient?

3. (a) Name the different machines or tools which turn woodboring bits. (b) Draw the one which is simplest to use.

4. Make a clear sketch to show how a woodboring bit is held in the chuck of a joiner's brace.

5. (a) Using simple sketches show the difference between 'countersinking' and 'counterboring'. (b) State clearly how and why this is done.

1.12 HOLDING WOOD

Tools are needed for holding wood when it is being worked to size and shape, or when work-pieces are being assembled. Without them woodworking would be much more difficult.

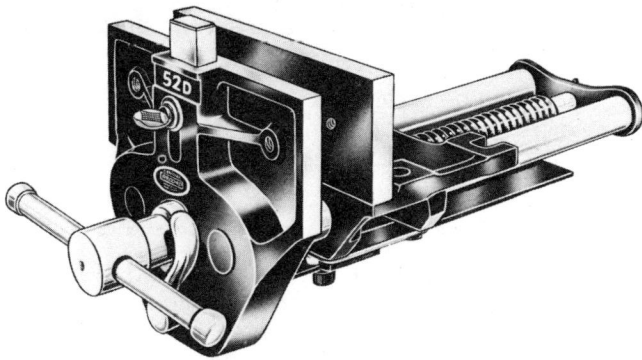

Fig. 77 Bench vice

Fig. 78 Sash cramp

Bench vice

Made from cast iron (Fig. 77) with accurately machined slide bars and screw of bright drawn steel. The vice can be used to hold wood when sawing, planing, chiselling, drilling, cleaning up, assembling, and within the limits of its capacity, for cramping up work after gluing.

Sash cramp

The head and slides (Fig. 78) are made from cast iron and the bar and screw from bright drawn steel. It is designed for the cramping of large section workpieces during assembling and gluing. They are used by cabinet makers, joiners and boatbuilders to draw joints tightly together. The cramp is left in position until the glue has set.

Cramp heads

With cramp heads (Fig. 79) a bar cramp of any length can be quickly and easily made from a piece of strong timber 25 mm thick.

G cramp

Best quality G cramps (Fig. 80) are drop forged from steel. They are used to hold workpieces down to the bench top during sawing, chiselling and shaping operations or for simple assembling and gluing. The swivel screw-head will adjust to bevelled surfaces.

Edging cramp

Of similar construction to the G cramp (Fig. 81), it is designed to clamp glued edging strips to straight or curved work, making nails, screws or sash cramps unnecessary.

Corner cramp

Sometimes called the mitre framing cramp (Fig. 82), its open design makes it possible to glue and pin joints which are under cramping pressure. It can be screwed directly to the bench, or to a block of wood to be held in the vice.

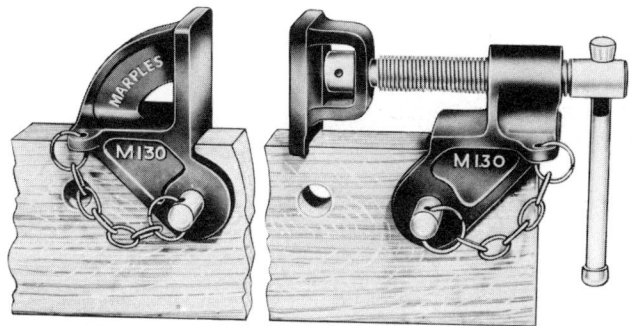

Fig. 79 Cramp heads

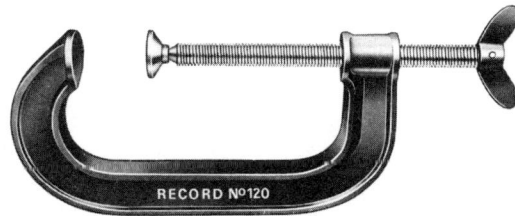

Fig. 80 G cramp

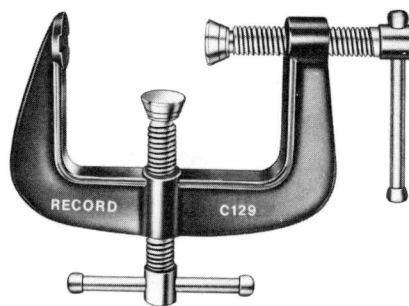

Fig. 81 Edging cramp

Fig. 82 Corner cramp

Fig. 83 Combined mitre cramp and sawing guide

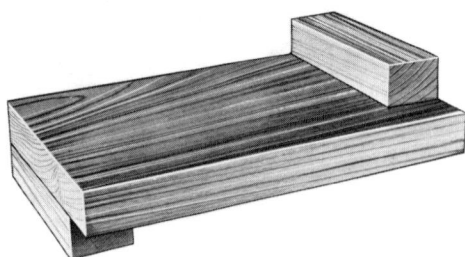

Fig. 84 Bench hook or sawing board

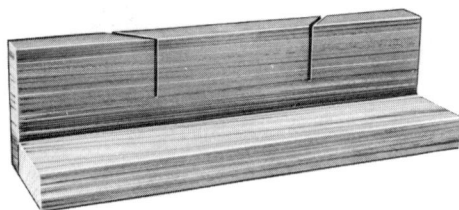

Fig. 85 Mitre block

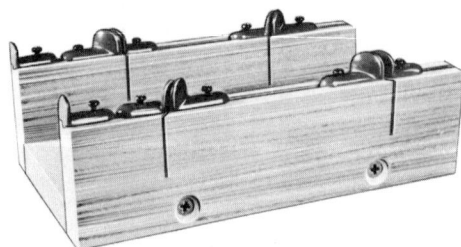

Fig. 86 Mitre box

Fig. 87 Woodcarver's screw

Combined mitre cramp and sawing guide

Of similar design (Fig. 83) to the corner cramp it can also be screwed to the bench. However the mitre sawing guide makes it less open, and so restricts nailing to some extent.

Bench hook or sawing board

A board (Fig. 84) on which to hold wood when it is being sawn with the tenon saw. Although it is not a saw guide it should be treated with great care to preserve its truth, to ensure accurate sawing. It can be held in the vice or against the front of the bench. The wood is then pressed on to it. The board prevents the saw from scoring the bench top.

Mitre block

This simple beechwood saw guide (Fig. 85) is used for cutting mitres on picture framing and mouldings. First make cut lines to saw on, and then lift the workpiece onto a strip of scrap wood level with the saw guides.

Mitre box

Will take mouldings up to 100 × 50 mm (Fig. 86). The adjustable saw guides fit snugly to the saw blade. To avoid damage to the set of the teeth do not push the saw down through them, or pull it up through them.

Woodcarver's screw

Holds carvings without obstruction (Fig. 87). Use the wing-nut as a wrench on the square shank to drive the screw into the wood. Drop the screw through hole in the bench, and tighten the nut underneath.

Dowelling jig

Once set, the jig (Fig. 88) can be inverted to drill the opposite set of holes to an equal depth.

Fig. 88 Dowelling jig

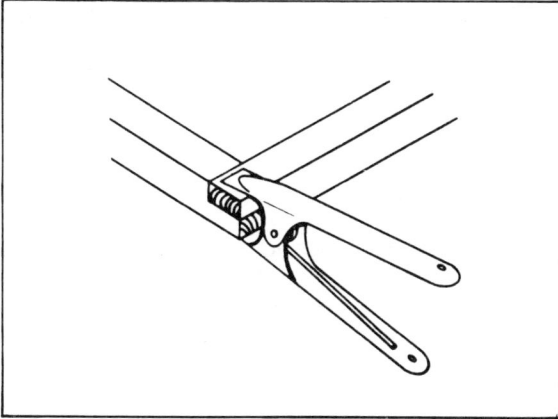
Fig. 89   Spring clamp

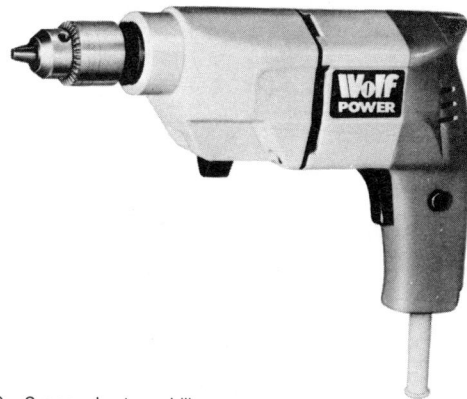
Fig. 90   2-speed rotary drill

Spring clamp

Enables frames and light components (Fig. 89) to be held firmly together during assembly, when being drilled or when small sections are being shaped by hand.

QUESTIONS

1. **(a)** Why are bench vices fitted with wooden cheeks and why must they be maintained in good condition? **(b)** What precautions can you take to ensure this?

2. Draw a G-cramp in use, showing how you would prevent the work from being damaged by it. Name the various parts.

3. **(a)** Draw a simple diagram of a piece of work for which sash cramps are needed when gluing up. Indicate the position of the cramps by arrows. **(b)** Are there any precautions you should take to ensure accuracy, and avoid damage to the work?

4. **(a)** Draw a mitre box. What is it used for? **(b)** Are there any precautions you should take when using it?

5. Show by a simple sectional sketch how a block of wood can be held to the bench top ready for carving.

1.13 ELECTRIC MACHINE TOOLS

Rotary drill

A choice of single and multispeed models (Fig. 90) with a range of chuck capacities is available to give the correct speed for drilling masonry, metal, timber or plastics. Used in a **drill stand** (Fig. 91) the machine is converted to a power tool capable of quite precise work. Its efficiency is further improved by the use of a **machine vice** (Fig. 92) in which to hold small work. When used in this way you must fit a **chuck guard** (Fig. 93) (H&S Act 1974).

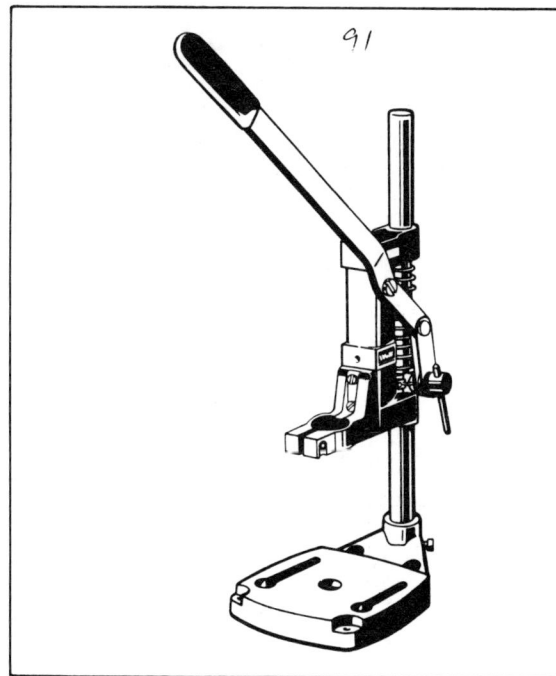
Fig. 91   Drill stand

Fig. 92   Machine vice

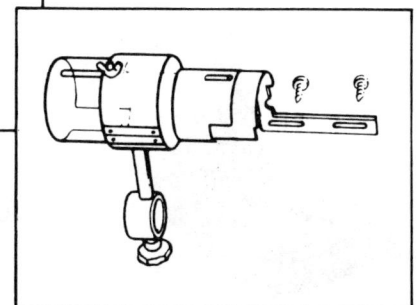
Fig. 93   Chuck guard

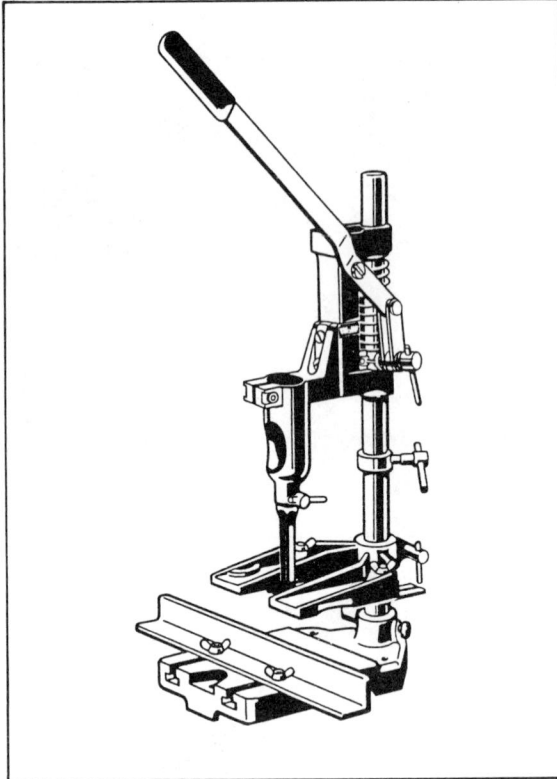
Fig. 94　Mortise drill stand

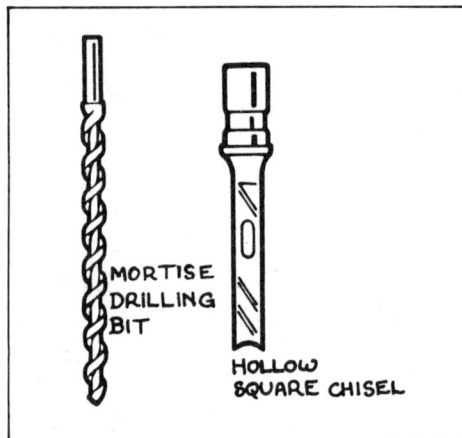
Fig. 95

The **mortise drill stand** (Fig. 94) is a further refinement enabling accurate mortises to be cut. These are made using mortise drilling bits and hollow square chisels (Fig. 95) of appropriate sizes.

Pedestal drill

This was originally designed to drill holes very accurately in metal (Fig. 96). Because of this, its construction is sturdy, making it a useful woodworking machine. As the need to fit 'knock-down' fittings with great accuracy has become essential, the value of the pedestal drill has further increased.

Jigsaw

The jigsaw (Fig. 97) comes as a 2-speed or variable speed model designed to cut either thin sheets or thicker boards. Straight cuts using a fence, or curved cuts, can be easily made in plywood, chipboard, blockboard, laminated plastics and thin metals.

Orbital sander

The pad of this fine-finishing sander (Fig. 98) makes only small orbits so avoiding marks on the grain. Graded self-adhesive, or clip-on sanding sheets should be used.

QUESTIONS

1. From your own experience in the workshop list and describe at least four uses for the hand-held rotary power drill.

2. Give reasons why a chuck guard is needed when using a vertical drilling machine.

3. (a) List the different ways of holding work securely to the drill table. (b) Draw simple diagrams and add explanatory notes.

Fig. 96　Pedestal drill

Fig. 97　2-speed jigsaw

Fig. 98　Orbital sander

1.14 WOOD TURNING

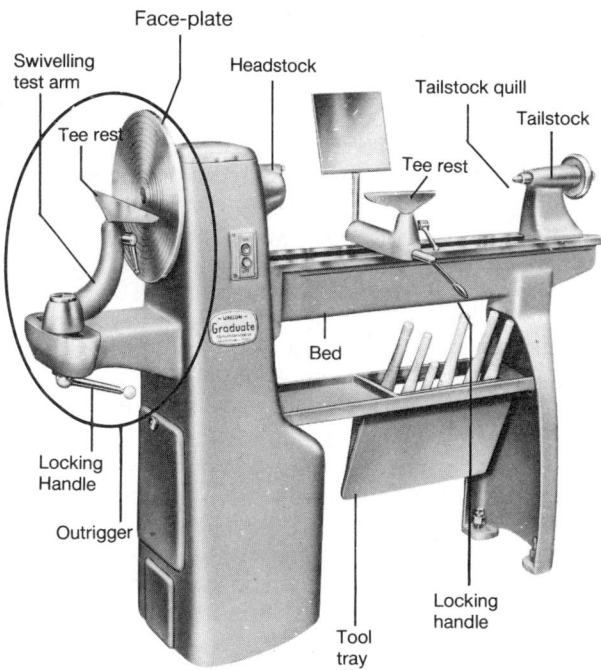

Fig. 99   Wood turning lathe

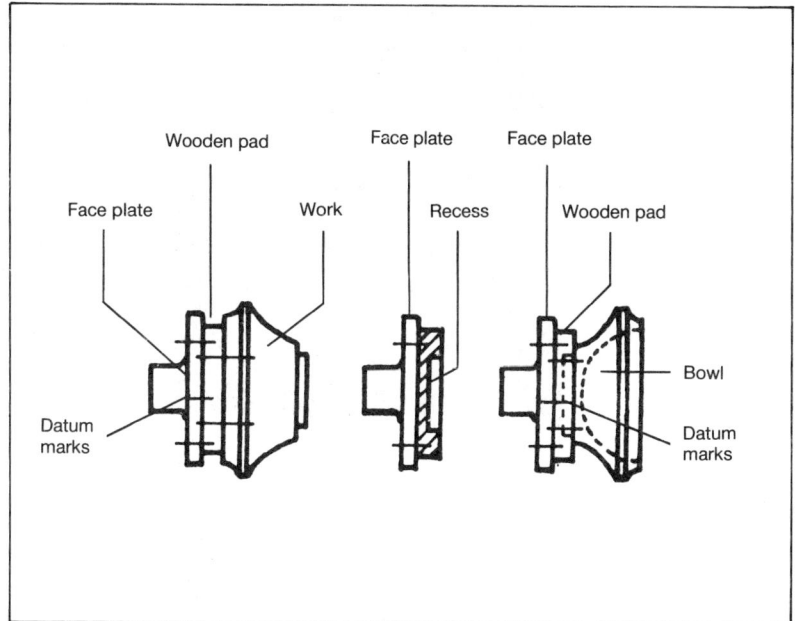

Fig. 100   Fig. 101   Fig. 102

Standard method of face-plate turning

a   One side of the blank for the work is planed flat, the waste corners sawn off and a circular wooden pad is screwed to the centre. The **face-plate** on which the pad was turned is then screwed back into place (Fig. 100). The purpose of the pad is to keep the turning tools clear of the metal face-plate so avoiding damage to their cutting edges as the outer profile and base of the bowl is turned to shape. Final smoothing is done with scraping tools and glasspaper of suitable grades.

b   A recess is now turned into the pad to accept the base of the bowl, and the two screwed together once more (Fig. 101). Light work can be glued to the pad with paper separating the two. They are then split apart when the work is complete.

c   The recess ensures the bowl runs true now it is reversed. Datum marks are required on the pad and face-plate so their correct positions are maintained (Fig. 102). The screws holding the bowl to the pad must not be too long or they will penetrate the bowl as the inside is turned out. This should be done firstly with the roughing gouge, the bull-nosed scraper and finally with an abrasive paper.

In Fig. 103 the direction of rotation is clockwise. Note the position of the gouge and also the recess in the base, into which a smaller face-plate or wooden pad can be fitted.

Some simple shapes can be screwed directly to the face-plate and the whole of the turning done in one operation.

Fig. 103   Turning a large diameter bowl

Turning between centres

Long work is turned between centres (Fig. 104).

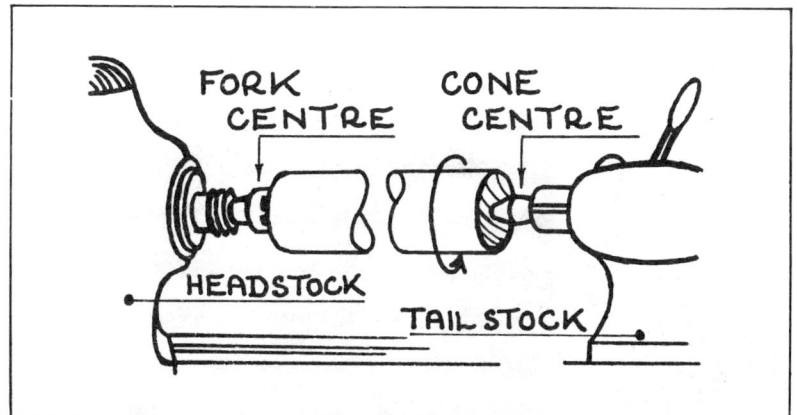

Fig. 104   Turning between centres

Fig. 105   Spinning wheel showing examples of turned work

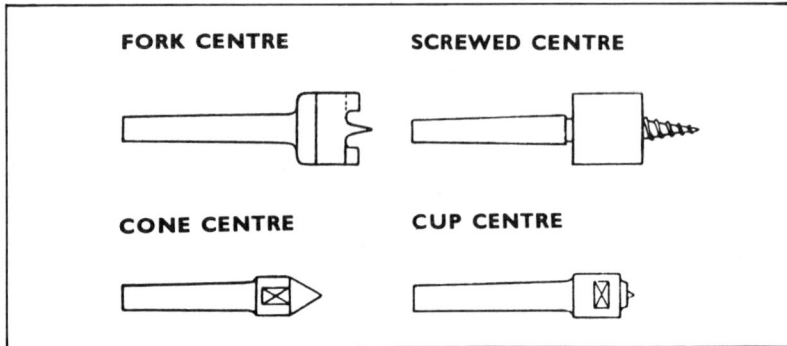

FORK CENTRE   **SCREWED CENTRE**

CONE CENTRE   **CUP CENTRE**

Fig. 106   Centres for turning wood

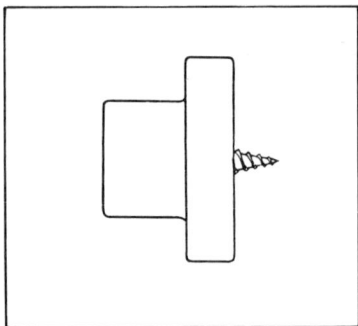

Fig. 107   Screw flange chuck

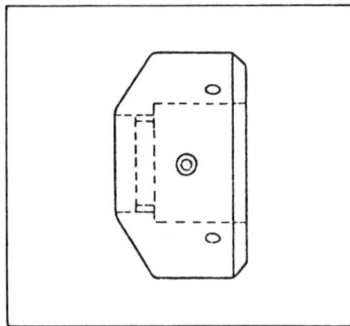

Fig. 108   Bell chuck

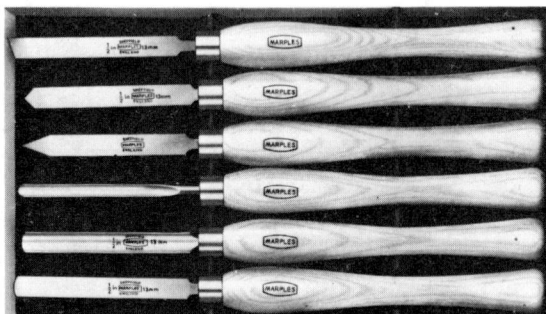

Fig. 109   Wood turning tools

Rolling pins, table and standard lamps, candlesticks and tool handles are typical examples (Fig. 105). To prepare the wood, centres are marked at each end by drawing the diagonals as this work is usually turned from square sections. From the centre a suitable circle is drawn and the corners planed off along the length of the wood. This reduces it to an octagonal section so there is less waste to turn off. Each centre is then dimpled with a punch, a bradawl or a small drill, depending on the type of work to be done. For example, small section work would not be punched because of the danger of splitting. Similarly, if the screw centre is to be used, a drilled hole is advisable, again to reduce the possibility of splits.

The **fork centre** (Fig. 106) is the one most frequently used to drive the wood round, and is fitted into a saw-cut made across one of the diagonals. The centre can be driven into the wood at this stage with a mallet, but on no account must a hammer be used or the morse taper will be damaged. The work can then be fitted to the lathe, the right-hand end being supported by a **cone centre**, or **cup centre** if it has to be deep bored, as in the case of a lamp stem.

Pressure on the work from the tailstock quill keeps it firmly in place between the two centres. The pressure is then maintained by tightening the locking screw to prevent any movement of the quill. To reduce friction and prevent overheating of these centres at the tailstock, the work should be lubricated with a non-staining substance such as tallow or soap. The Tee rest should then be positioned as close to the work as possible and at a suitable height, so the cutting edge of the turning tool is level with the centre of the wood.

Turning small bowls

Small bowls, discs and platters can be turned by mounting the wood on the '**screw flange' chuck** (Fig. 107). The work tightens onto the screw because of the pressure of the cutting tool. The '**bell' chuck** (Fig. 108) can be used as an alternative, to hold short work on which a strong boss has to be turned to fit into the bell, where it is held by an Allen screw. Both screw onto the lathe spindle, and if necessary the work can be given extra support with the cone centre.

Wood turning tools

Scraping tools should be held with the tips downwards to give the best results where there is difficult grain. Where possible, other cuts should always be with the grain.

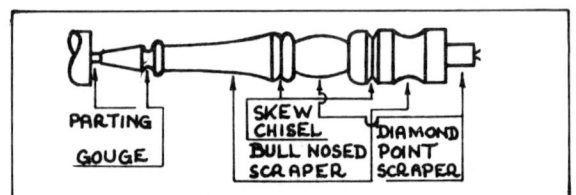

PARTING   SKEW
            CHISEL   DIAMOND
GOUGE   BULL NOSED POINT
            SCRAPER   SCRAPER

Fig. 110   Common wood turning cuts

QUESTIONS

1. (a) Sketch the cutting edges of common wood turning tools including a short length of the blade. (b) How are they sharpened?

2. The Tee rest should always be close to the work being turned. At what height should it be for the various tool operations? Make simple sketches.

3. The fork centre is often used when long pieces are being turned. (a) Draw a fork centre showing all its details. (b) Make sketches of the other centres which can support the other end of the wood.

4. (a) Draw sketches of how work may be mounted for face-plate turning. (b) Indicate any precautions you must take to avoid damage to the work or turning tools.

5. (a) Where and when would you use a scraping tool? (b) How are these sharpened? (c) How are they held to the work? Draw sketches to illustrate your answer.

REVISION QUESTIONS

Measuring, setting out, testing and gauges

1. Tools such as the try square, mitre square, sliding bevel and gauges, should preferably always be held against the face side or face edge of the wood. (a) Explain carefully why this is so. (b) How is waste wood marked, and why is this advisable? Make simple sketches to illustrate your answers.

2. Make sketches to show how you would (a) set a single marking gauge to 100 mm to mark off the width of a piece of wood (b) set a mortise gauge to mark out a mortise in the centre of a door stile. State clearly how each is adjusted.

Saws

3. Which saw would you choose to cut a very dense piece of chipboard and say why.

4. How, and with which saw would you:
(a) cut along the length of a wide board.
(b) cut out a piece of 4 mm plywood from a large sheet.
(c) cut off the waste from a hardwood rail 50 × 20 mm in section.
(d) cut the dovetails at the corners of a small box.
(e) cut out a ∅300 circle for a speaker box made from 16 mm chipboard.
(f) cut out the finger grip in the handle of a cutlery box? Give reasons for your choice.

5. (a) List the saws used for cutting wood. (b) What precautions should you take when using each one?

Abrasive tools and planes

6. (a) List the ways in which splintering of the wood may occur when using abrasive tools such as rasps and Surforms? (b) How can it be avoided?

7. (a) What do you understand by the term 'frog'? (b) Why is it adjustable on metal planes?

8. (a) What are the advantages of the corrugated sole on bench planes? (b) What are the disadvantages?

9. (a) Which special plane would you use to:
(i) prepare the framework for a small door which is to be glazed.
(ii) prepare a framework to hold a wooden panel.
(iii) true up the surface of a stopped chamfer.
(iv) plane the end grain of a small piece of very hard wood?
(b) What features do the blades of each of these planes have in common?

10. Circular planes and spokeshaves can be used for similar work. Give examples of when you would use one in preference to the other.

11. (a) What does 'planing with the grain' mean? (b) What special precautions must you take when planing end grain and how is this done? Illustrate your answer with simple sketches.

Scraping tools, chisels and gouges

12. (a) Draw your own design for a scratch stock. (b) Make sketches to show the profile of a blade suitable for scratching out (i) small double flutes (ii) small double reeds in hardwood framing.

13. (a) Sketch the different ways a chisel handle can be fitted to the blade. (b) Why are the different methods used? (c) What are the special features of the different handles?

14. (a) Why must the back of a chisel blade be perfectly flat at all times? (b) What advice would you give on the use and care of chisels? Write from your own experience.

15. Draw the grinding and sharpening bevels of the two common workshop gouges.

16. Design a piece of work which involves the use of the 'in-cannel' and the 'out-cannel' gouge.

17. (a) Using sketches show how you would carve your initials on the top of a hardwood box. (b) Illustrate the tools you would use.

Hammers, mallets and screwdrivers

18. Illustrate the different ways heads of mallets and hammers are fitted and fixed to the shaft. Add notes to explain your sketches.

19. Describe how you should take care of screwdrivers.

20. Using sketches and notes distinguish between a cabinet screwdriver and a London pattern screwdriver.

21. A screwdriver blade can slip from a screw head and damage the wood. In what circumstances might this happen?

22. Discuss the advantages and disadvantages of the square-bladed 'birdcage' bradawl and the chisel-pointed bradawl.

23. (a) Draw a screw start. (b) Describe three different occasions when you would use it.

Wood boring

24. Sketch examples of work where you would use each of the three auger bits.

25. If a wood boring bit bursts through the wood it can cause a lot of splintering. Make sketches and notes to show how this can be avoided.

26. When would you use a hand drill instead of a brace and bit? State your reasons clearly.

Holding tools

27. (**a**) List six different tool operations carried out with the help of a vice. (**b**) Sketch the tools you would use for each operation?

28. When would you use (**a**) G-cramps (**b**) edging cramps (**c**) mitre cramps? Include sketches; the cramps need not be shown, but indicate their position with arrows.

29. Using step-by-step sketches, show how you would make a bench hook.

30. (**a**) List, in order, all the tools you would use to set out mitres, before cutting them on the mitre block or in the mitre box. (**b**) Distinguish between the mitre block and the mitre box.

31. Sketch a simple dowelling jig and say how it is used.

Electric machine tools

32. Not all power drills have the same specifications. Find out (**a**) what the different specifications are (**b**) how and why they differ.

33. Design a piece of work to be made entirely on the pedestal drilling machine. Any combination of machine drills may be used.

34. (**a**) What is 'plunge-cutting'? (**b**) List the different uses of the jig-saw. Illustrate with simple sketches.

35. (**a**) What is an 'orbital sander'? (**b**) In finishing a piece of work what are the stages to be worked through before the sander is used?

36. (**a**) What does the term 'double-insulation' mean? (**b**) What special precautions should be taken when using electric hand tools?

Wood turning lathe

37. Write down the names of the different parts of the wood turning lathe and describe briefly the function of each one.

38. The basic function of the lathe is to make wood cylindrical (circular in section). Describe briefly the full range of work which can be made on it. Make sketches if you wish.

39. (**a**) Describe how you would smooth and finally finish a wooden bowl. (**b**) Where are the difficult parts likely to be, and why? Illustrate by means of a simple sketch. (**c**) What are the alternative finishes you could use?

40. (**a**) What is the purpose of the lathe outrigger? (**b**) In use how is the Tee rest set? (**c**) How are the cutting tools used?

TWO

Timber

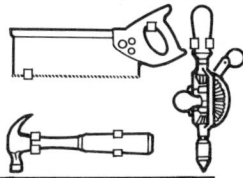

2.1 TREE GROWTH AND PARTS OF THE TREE

Tree growth

The manufacture of food vital to the life of the tree takes place in its leaves (Fig. 111). The food is a compound of carbon and a kind of sugar, the carbon being derived from carbon dioxide in the atmosphere. Water, the other vital ingredient, is carried to the leaves from the roots (Fig. 112) by the sapwood. Chlorophyll, the green pigment of the leaf, needed for this process, is activated by the radiant energy of the sun. By photosynthesis the carbon and water are changed to sugar and oxygen. Excess water is given off through the stoma by transpiration. These pores close at night and open wide during the day particularly in bright sunlight. As photosynthesis takes place oxygen is released into the atmosphere.

Fig. 112   Part of the root system of a fallen beech

Parts of the tree

Bark protects the tree, growing as the tree increases in size.
Cambium where new wood grows inwardly and the bast outwardly.
Bast through which food made by the leaves is carried down to branches, trunk and roots.
Sapwood sap passes through this from the roots to the leaves.
Heartwood inactive, but gives strength to the trunk, and is where the best timber is obtained.
Medullary rays food is stored by the rays and dispersed horizontally. Generally they are much smaller in softwoods and therefore more difficult to see than the pronounced medullary rays of many hardwoods (Fig. 114 – See over). Medullary

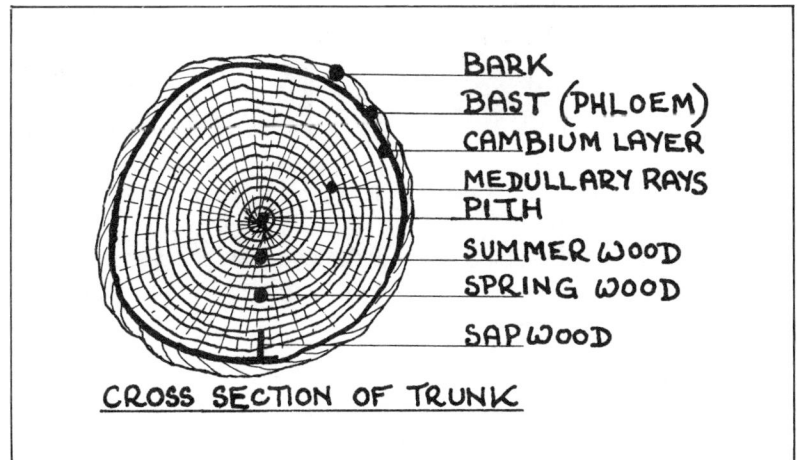

BARK
BAST (PHLOEM)
CAMBIUM LAYER
MEDULLARY RAYS
PITH
SUMMER WOOD
SPRING WOOD
SAPWOOD

CROSS SECTION OF TRUNK

Fig. 113

Fig. 111   Section of a leaf

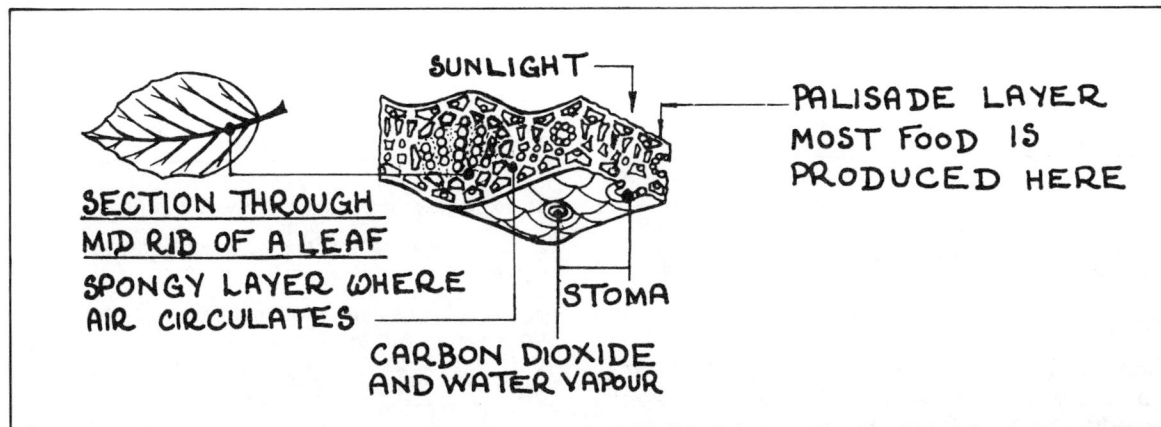

SUNLIGHT
PALISADE LAYER MOST FOOD IS PRODUCED HERE
SECTION THROUGH MID RIB OF A LEAF
SPONGY LAYER WHERE AIR CIRCULATES
STOMA
CARBON DIOXIDE AND WATER VAPOUR

Fig. 114 Medullary rays and annual rings visible in this cross-section of opalized oak

Fig. 115 Section of an oak fork showing how the grain gives the branch its strength at the junction with the main stem

rays bind together the layers of spring and summer wood. These are often referred to as the **annual rings**.

QUESTIONS

1. How does a tree obtain its food from the earth?

2. What is the function of the leaves?

3. Draw the cross-section of a hardwood tree and name the various parts.

4. (a) Where does the most active growth occur?
(b) Draw an enlarged section of this part.

2.2 SOFTWOODS AND HARDWOODS

Softwoods

Conifers (Fig. 116) with their straight trunks, coarse bark and small branches produce softwoods. These trees are important to the world's economy because they are fast growing even in poor soils and in harsh climates. The timber they produce is very easy to work. They have a simple structure and growth pattern which varies very little between species.

Conifers are evergreen (except for the larch) with narrow and needle-like or scale-like leaves (Fig. 117). Another distinctive feature is the woody cone or fruit, in which the seed develops. Their branches are regular, almost geometrical, while the trees themselves have a resinous smell.

Hardwoods

The hardwood trees, or broadleaves are part of the natural forest of Northern Europe. The broad leaf is the principal feature of all these trees which are deciduous, the leaf being shed each autumn as winter approaches. Some of the most common are featured below. Hardwood trees are also native to North America and the tropical rain forests.

Oak grows in the temperate forests of the Northern Hemisphere in Europe, Asia and America (Fig. 118) and can be easily identified by their familiar leaves and acorns (Fig. 119).

Ash tall upstanding trees with a wide-spreading crown (Fig. 120). Their branches tend to fork readily.

Sycamore is a typical maple common to the forests of Central Europe, and is now firmly established in our woodlands (Fig. 122).

Beech its distinctive form is easily identifiable (Fig. 124) and difficult to confuse with any other type of tree.

Elm the English Elm (Fig. 126) displays its typical form.

In general hardwoods are heavier, darker in colour, and much harder than softwoods. Hardwoods grown in the Northern Hemisphere have a distinct colour and grain pattern, caused by the contrasting annual rings of spring and summer growth. In the tropical regions where growing seasons are not so well defined, new wood is formed in concentric layers which may not be so distinct, or annual.

QUESTIONS

1. Explain what the term 'softwood' means.

2. List five characteristics by which softwood trees can be identified.

3. How does a hardwood tree differ in appearance from a softwood tree?

4. In what ways do hardwoods differ from softwoods?

Fig. 116   Conifer plantation

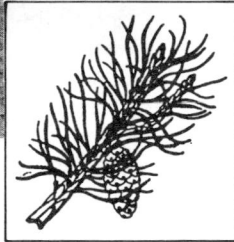

Fig. 117   Needle-like leaves and cone

Fig. 118   English Oak

Fig. 119   Oak leaf and acorns

Fig. 120   English Ash

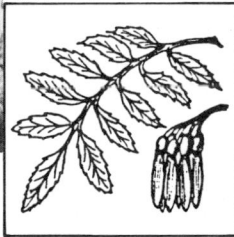

Fig. 121   Ash leaf and seeds

Fig. 122   European Sycamore

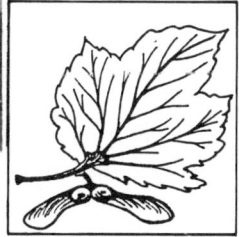

Fig. 123   Sycamore leaf and seeds

Fig. 124   European Beech

Fig. 125   Beech leaf and fruit

Fig. 126   English Elm

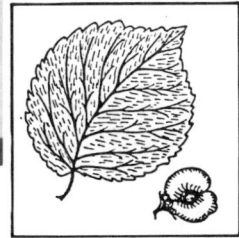

Fig. 127   Elm leaf and seeds

Fig. 128   Cutting boards from a log

Fig. 129   Vertical band saw for first stage conversion of a hardwood log

2.3 CONVERSION

Logs were originally used in their natural state or roughly hewn to shape with axes or an adze. Boards or thinner pieces were made from them by splitting along the grain with wedges. With the growing need for more precise sections, and more accurate work, boards were sawn from the log by hand (Fig. 128). This was done between two trestles, or across a pit using a very long, double-handed saw. This was, of course, a slow and arduous task, but one which later became easier with the introduction of machines.

Both circular saws, and band saws were developed. Heavy-duty types are now used in converting logs into commercial sections of timber (Fig. 129). Conversion of the log at the saw mill is known as '**breaking down**' for which '**spring set**' saws are used to cut 'across the grain', i.e. cross cutting, or '**swage set**' to rip along the grain (Fig. 130). Conversion may be by simple '**through and through**' cuts (Fig. 131) or by '**quartering**' (Fig. 132). Quartering can be done in several ways, four of which are shown. The log is cut parallel or near to the medullary rays exposing them to provide better figured timber, which has more stability and greater resistance to wear. '**Tangential**' sawing (Fig. 133) is used to show off the pronounced grain of certain woods such as pitch pine.

Newly felled 'green' timber should be converted as soon as possible, because evaporation of the wood fluids along the log, and particularly at the ends will cause splits as the timber dries out and contracts. This could result in much valuable timber being wasted.

QUESTIONS

1. What does the term 'conversion of timber' mean?

2. Using sketches distinguish between (**a**) plain sawn, (**b**) quarter sawn, (**c**) tangential sawn. Why are the different cuts used?

3. Write a short account of the improvements made in the methods of converting logs into timber.

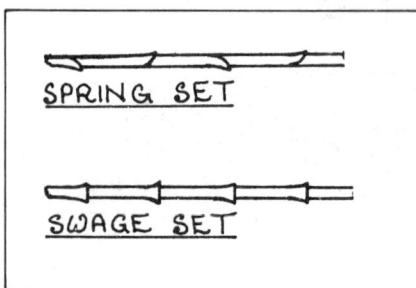

SPRING SET

SWAGE SET

Fig. 130

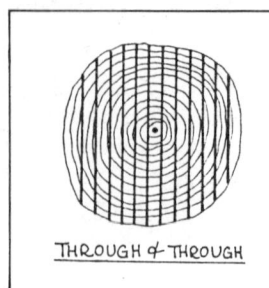

THROUGH & THROUGH

Fig. 131

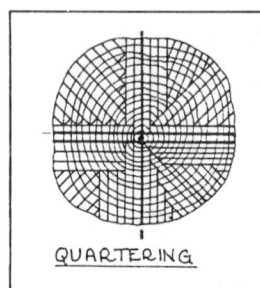

QUARTERING

Fig. 132

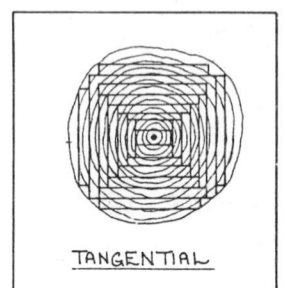

TANGENTIAL

Fig. 133

2.4 SEASONING AND MOISTURE CONTENT

Once the timber has been converted, the sawn boards are then dried out by '**seasoning**' to reduce the moisture content to an acceptable level, depending upon what the timber is to be used for. The '**moisture content**' should always be less than 20%, which is the weight of the moisture expressed as a percentage of the weight of the timber when seasoned. Therefore a moisture content of 20% means that for every 1 kg of moisture there are 5 kg of timber.

Natural seasoning (Fig. 134) by air drying takes many years and the timber must be stacked correctly, protected from the weather, but in such a way that the air can circulate freely between each piece.

Kiln seasoning (Fig. 135) takes only a few days by comparison and the moisture content can be predetermined.

In each case sticks between the timber must be of equal thickness and all directly above each other or there will be undue distortion (Fig. 136).

The moisture content of the timber should be similar to the humidity of its surroundings. See below.

20–16% exterior joinery, exterior doors, garages and sheds, garden furniture, fencing etc.

16–13% interior joinery, bedroom furniture, interior doors, timber where there is occasional heating.

13–10% wood in buildings where there is continuous heating.

10–8% wood close to regular heat i.e. block floors and skirting boards over heating elements.

QUESTIONS

1. (**a**) Why is timber 'seasoned'? (**b**) What are the methods by which this may be done? Draw simple illustrations.

2. The log of a newly felled tree has been plain sawn into 40 mm boards. (**a**) How would it be stacked for natural air seasoning? (**b**) What precautions are necessary to avoid defects occurring?

3. What do you understand by the term 'moisture content'?

2.5 DEFECTS

Seasoned timber has increased stability, greater strength and more resistance to decay, but during seasoning the annual rings shorten. This causes

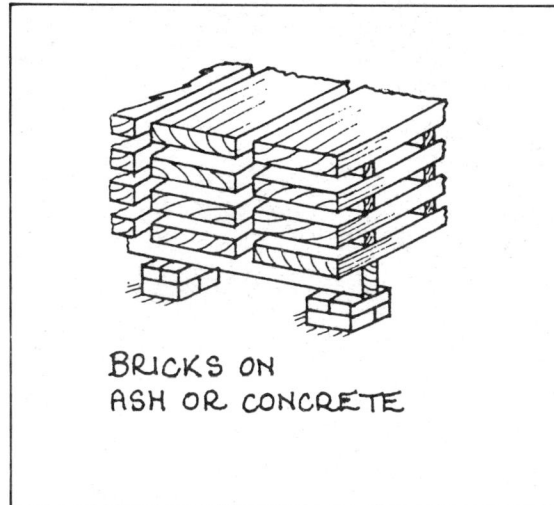

Fig. 134   Natural seasoning

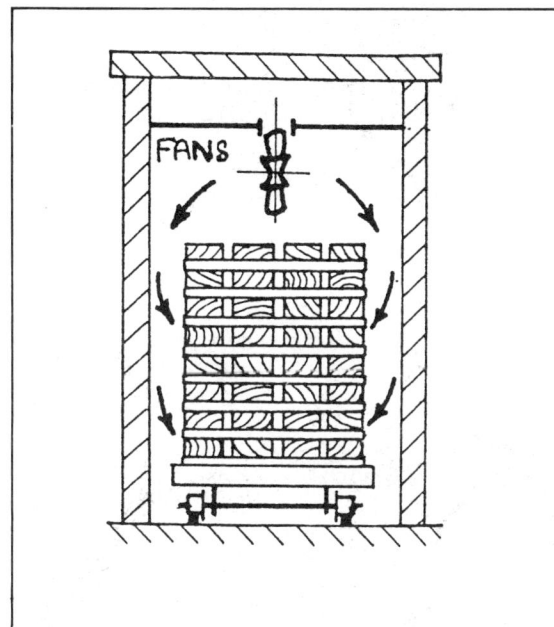

Fig. 135   Kiln seasoning by careful control of hot air and steam

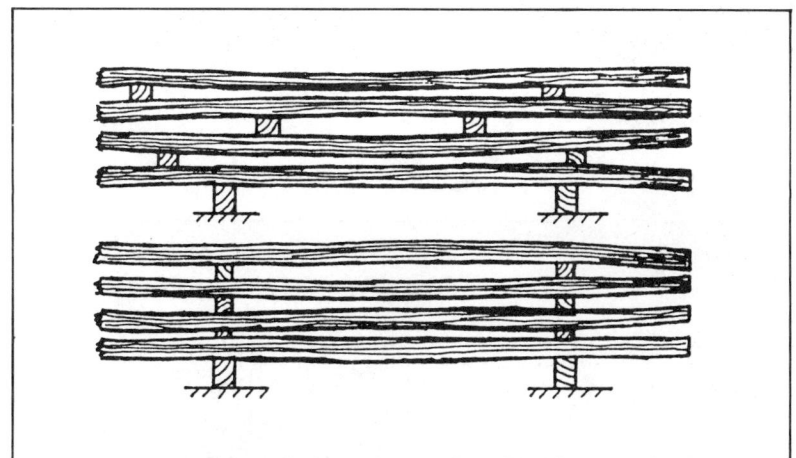

Fig. 136   Examples of incorrect stacking

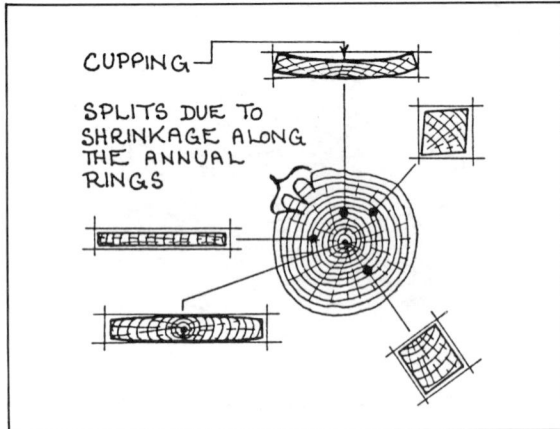
Fig. 137   Tangential shrinkage

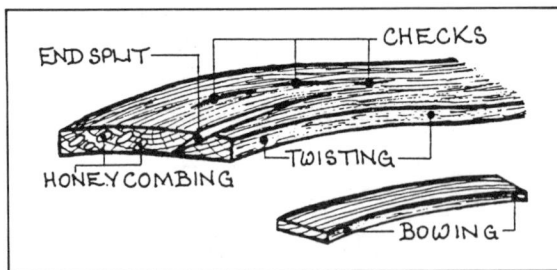
Fig. 138   Seasoning defects

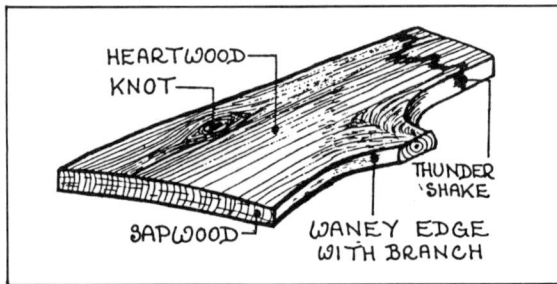
Fig. 139   Natural defects

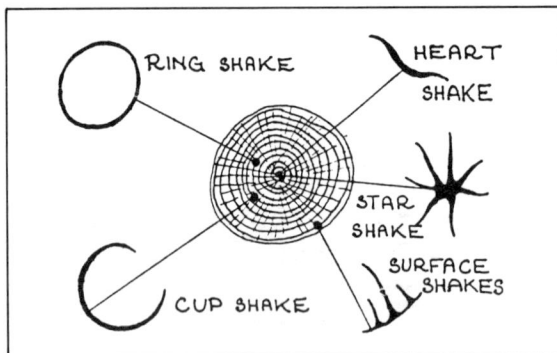
Fig. 140

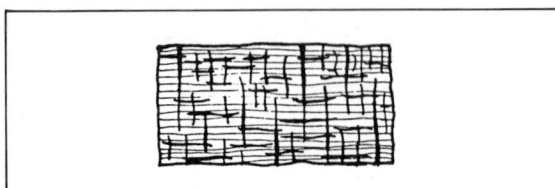
Fig. 141   Effect of dry rot

some distortion of the sectional shape, called '**tangential shrinkage**'. The diagram (Fig. 137) indicates how this varies, depending on where the timber is cut from the baulk.

Other defects (Fig. 138) which may occur during seasoning are:
End splits cracks following the grain.
Honey comb small open pockets within the board.
Checks small surface cracks where the fibres have parted.
Twisting spiral distortion.
Bowing curving along the length.

Natural defects (Fig. 139) are generally well known. **Shakes** (Fig. 140) which are splits in the wood, may occur when the tree is standing, during felling or before conversion and may result in considerable wastage. Shakes are found round the annual rings or along the medullary rays.

Knots are caused by the growth of a branch from the main stem. They often cause severe difficulties in working the wood, and considerable variations in the grain. '**Dead Knots**' are those which are loose and liable to drop out. Resin ducts are pockets of aromatic but sticky resin usually associated with coniferous woods.

QUESTIONS

1. (**a**) What causes 'tangential shrinkage'? (**b**) What effect does it have on the sectional shape of plain sawn boards?

2. Make sketches to show how shortening of the annual rings during seasoning causes distortion of the sectional shape of timber depending on where it has been cut from the baulk.

3. Using notes and sketches explain the following terms (**a**) star shake, (**b**) cup shake, (**c**) ring shake, (**d**) heart shake, (**e**) twisting and bowing (warping), (**f**) waney edge.

4. (**a**) What causes knots in timber? (**b**) What is a 'dead knot'? (**c**) What is a resin duct and in which timbers is it likely to be found?

2.6 DECAY AND INSECT ATTACK

Timber in badly ventilated situations, and where the moisture content becomes more than 20% is susceptible to fungal attack, in the form of wet rot or dry rot. The most serious of these is **dry rot** (Fig. 141). The name refers to the decayed timber and not the conditions for its growth. Infected timber becomes very dry and brittle and can be crumbled in the hand. Long cracks across the grain are a distinctive feature and there is a pronounced musty odour in badly affected areas.

Eradication can be difficult and is best carried out by specialists.

There are four beetles (Fig. 142) whose larvae attack timber. They are:

A The Furniture Beetle—the most common.
B The Powder Post
C The Death Watch
D The House Longhorn

Insecticides applied by brush, spray, or by injection into the flight holes killing the larvae, are very effective.

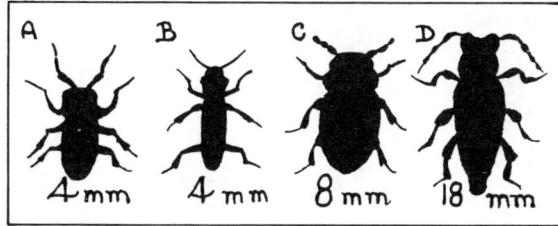

Fig. 142

QUESTIONS

1. (**a**) Under what circumstances would fungii attack timber? (**b**) Name the different species.

2. (**a**) Which is the most serious of the above fungii? (**b**) How would you identify it?

3. Name the insects which attack timber and draw sketches approximately three times full-size of each one.

4. How would you recognise and treat timber affected by insect attack?

Fig. 143   Garden workshop made from Red Deal

2.7 TIMBER DATA SOFTWOODS

Timber is now sold in metric measure being cut at saw mills to standard lengths and sections. Lengths are measured in **Metres** (m) and sections are measured in **Millimetres** (mm).

Standard widths begin at 75 mm increasing to 300 mm and thicknesses range from 12 mm to 300 mm (Fig. 144). There are nine standard widths and sixteen standard thicknesses. These are the thicknesses to which timber is cut at the saw mill and are called **nominal**.

Often the timber merchant supplies his customers with planed timber so the thickness is reduced considerably. There is a corresponding reduction in width if the edges are also planed i.e. ' **planed all round**' (p.a.r.). Alternative abbreviations are 'planed square edged' (p.s.e.) or 'planed and square jointed' (p.s.j.).

The chart shows **sawn** (nominal) thicknesses. When '**planed two sides**' (P2S) they are considerably thinner, e.g. 12 becomes 8, 16 becomes 12 and so on. Small sections are re-sawn from standard nominal sizes of 25 × 25, 25 × 38, 25 × 50, and 50 × 50 mm.

Sawn softwood

The standard range of lengths supplied from the mills (Fig. 145) begins at 1.8 m increasing in steps of 300 mm to 6.3 m. Shorter and longer lengths can be obtained to order.

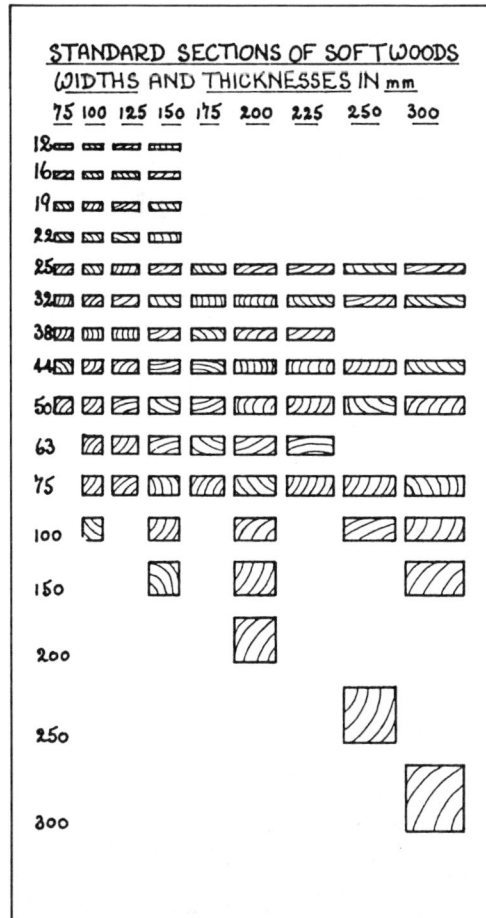

Fig. 144

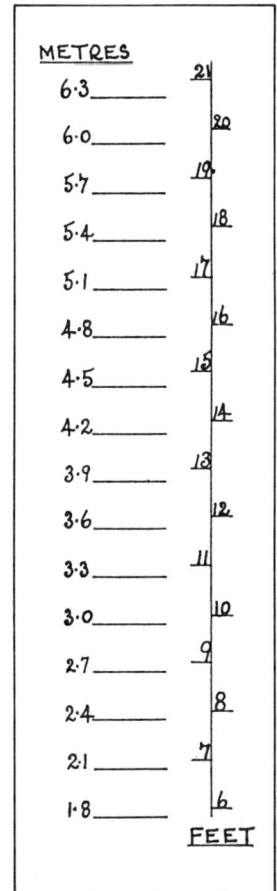

Fig. 145   Standard lengths of sawn softwood

Fig. 146   Furniture being constructed from hardwood

QUESTIONS

1. (**a**) When buying timber in what units are the lengths and section specified? (**b**) Give three examples for softwoods.

2. (**a**) Make two lists to show the full range of standard widths and thicknesses of softwoods. (**b**) How many of each are there?

3. (**a**) What does the term 'nominal' mean? (**b**) If you order softwood 'planed all round' how does this affect the standard width and thickness? Give examples. (**c**) What does P2S mean?

4. (**a**) Make a list to show the standard metric lengths of sawn softwoods. (**b**) Are non-standard lengths available? If so how are they obtained?

2.8 TIMBER DATA HARDWOODS

The thicknesses of sawn hardwood are similar to those for softwood up to 125 mm, then rising in stages of 25 mm. The thicknesses are 12, 16, 19, 25, 32, 38, 50, 63, 75, 100 and 125 mm.

Hardwood boards are available in widths of 150 mm and upwards in steps of 10 mm. Because the lengths and widths of hardwood boards do vary considerably, being cut to avoid unnecessary waste, they are usually sold by the square metre.

Sawn hardwood

The standard range of lengths supplied from the mills begins at 1.8 m but increase in steps of 100 mm (Fig. 147). Some hardwoods are imported in lengths shorter than 1.8 m.

The metric units of sale are:
By cross section in millimetres
Lineal metres (m)
Square metres (m^2 or sq. m)
Cubic metres (m^3 or cu. m)

METRES

METRES	FEET	INCHES
6·3	20	8
6·0	19	8$\frac{1}{8}$
5·7	18	8$\frac{3}{8}$
5·4	17	8$\frac{5}{8}$
5·1	16	8$\frac{3}{4}$
4·8	15	9
4·5	14	9$\frac{1}{8}$
4·2	13	9$\frac{3}{8}$
3·9	12	9$\frac{1}{2}$
3·6	11	9$\frac{3}{4}$
3·3	10	9$\frac{7}{8}$
3·0	9	10$\frac{1}{8}$
2·7	8	10$\frac{1}{4}$
2·4	7	10$\frac{1}{2}$
2·1	6	10$\frac{5}{8}$
1·8	5	10$\frac{7}{8}$

Fig. 147   Standard lengths of sawn hardwoods

QUESTIONS

1. List the thicknesses of sawn hardwood boards.

2. What are the standard widths of hardwood boards?

3. In what way do the thicknesses of sawn hardwoods vary from those of sawn softwoods?

4. Why are hardwood boards usually sold by the square metre?

5. (**a**) Why does the range for standard lengths of sawn hardwood increase in steps of only 100 mm? (**b**) From which part of the tree is the best hardwood obtained? Give the reasons for your answer.

2.9 SOFTWOODS TABLE

COMMON NAME	WORLD SOURCE	WORKING QUALITIES	USES	CHARACTERISTICS— GRAIN, WEIGHT, COLOUR ETC.
Red Deal or Redwood Scots Pine Baltic Pine	Northern Europe, Scandinavia and Russia, Spain, Scotland and parts of England and Wales.	Easily worked—takes nails and screws readily and glues well. Varnishes and paints well when correctly primed.	Durable as telegraph poles, railway sleepers, fencing when creosoted or treated with preservative. Leading building timber for joists, rafters, flooring, sheds, pit-props, etc. and general domestic purposes.	May be knotty, some knots are large but usually firm. Fine to coarse grain. Medium weight. Red-brown heartwood. Creamy-fawn sapwood.
Larch	Europe and Asia.	Saws well, takes nails and other fastenings readily. Very sharp tools required for satisfactory working.	Outdoor carpentry because of its durability. Estate repair work— fencing, gates, garden furniture etc.	Clean, long grain. Durable with greater strength than most conifers. Medium-heavy. Red-brown heartwood. Creamy-brown sapwood.
Whitewood Sitka and Norway Spruce	Western North America, Canada and Europe.	Difficult to work because it is soft. Requires sharp tools to produce a good finish. Readily accepts nails, screws, glue and paint.	Shelving and interior joinery, packing cases for food (does not impart any smell to contents). Chipboard and wood pulp for paper making.	Even texture—wide annual rings. Light in weight. Pale creamy brown to almost white. Little colour difference between heartwood and softwood.
Yew	British Isles and Europe.	Hard to work with common handtools. Care needed if nails or screws are to be used. Accepts glues.	Decorative furniture, bowls and ornaments. Formerly used for long bows.	Often 'wild' lustrous grain. Heavy weight. Heartwood red to purple. Creamy sapwood.
Oregon or Columbian Pine Douglas Fir	West coast of North America, Canada, Europe and Scandinavia.	Hard timber, quite difficult to work. Takes a good finish, water resistant and very stable. Inclined to split when nailed; accepts screws and glue readily.	Advantage is taken of its attractive grain for interior woodwork. Heavy constructional work—gates, bridges, poles and masts. Modern furniture, heavy-duty plywoods.	Generally straight even grain. Almost knot free 'clear' boards. Spring and summer growth clearly distinguished. Medium-heavy. Orange-brown heartwood. Creamy-brown sapwood.
Western Hemlock	West coast of North America, Canada and Asia.	Rather soft working, requires very sharp tools. Accepts glues, nails and screws with ease.	Exterior doors, gutters. Interior joinery and decorative panelling because of its attractive grain. Box making.	Fine, uniform texture. Available in considerable widths. Medium weight. Pale brown colour.
Western Red Cedar	Canada and Western North America.	Easy to work to a fine, silky surface with sharp tools.	Interior, and exterior joinery for which it is very suitable because of its resistance to fungal decay. Not suitable for heavy duty work as it lacks structural strength.	Close, even, straight grain. Large boards free from knots and defects. Light-medium weight. Orange-brown weathering to silvery-grey.
Cypress (foliage and growth very similar to Western Red Cedar)	Cyprus, Asia Minor, China, Himalayas.	Works similarly to larch, but little is grown for commercial purposes.	A most popular tree for garden hedges and ornamental planting.	Fast growing so spring and summer growths are easily distinguished. Medium weight. Cream to fawn colour.

QUESTIONS

1. List the softwoods named in the table and beside each, write down the working qualities they have in common.

2. Give the alternative names of Red Deal and identify its main characteristics and uses.

3. Contrast the characteristics of Larch, Spruce and Douglas Fir.

4. List the main uses of Red Deal, Larch, Spruce, Douglas Fir and Western Hemlock.

5. (a) Why are most softwoods available in long sections and wide boards? Name the exceptions to this. (b) Why do most softwoods readily accept nails, screws and other fastenings?

6. List the softwoods you would choose to make the following items, and give the reasons for your choice. (a) garden sun house (b) a room divider (c) a garden bench seat (d) a table lamp.

2.10 HARDWOODS TABLE

COMMON NAME	WORLD SOURCE	WORKING QUALITIES	USES	CHARACTERISTICS– GRAIN, WEIGHT, COLOUR ETC.
English Oak	Great Britain.	Strong and hard. Very difficult to nail. Can be secured with pegs and screws which should be of brass to avoid staining. Steel screws cause iron stains because of the tannic acid in Oak. Takes a good finish.	Best quality joinery where maximum durability is required. Furniture of all kinds. Ladder rungs, wheel spokes and barrel staves. Riven or cleft oak is excellent for fencing posts and rails.	Conspicuous grain, annual rings and medullary rays easily traced. Heavy. Heartwood rich, deep brown, sapwood cream to fawn. The heartwood is very durable and resistant to insect and fungal attack.
Ash	Europe, North America.	Variable—may be quite mild working, or very tough and springy. Tools always need to be very sharp. Difficult to nail or screw but glues well.	Used for strong handles for hammers, axes, spades etc. Hockey-sticks, oars, the rims of wooden wheels and coach building. Special furniture, because it can be easily bent to curved outlines by steaming.	Grain generally long and straight giving the timber its characteristic elasticity. Hard in texture with very open grain. Heavy. Generally yellow-white, the heartwood of older trees often deep brown.
Sycamore	Europe.	Hard. Very sharp tools required with planes set fine. Pilot holes required if nails or screws are to be used. Generally has a straight grain giving a good finish, but if irregular is difficult to work.	Domestic woodware, draining boards, cutting boards, etc. Turnery.	Fine, close grain. Growth rings not clearly defined. Heavy weight. Pale cream or white in colour throughout. In old trees the heartwood may become brown. Water can produce blue-grey stains.
Beech	Europe.	Hard and difficult to work but sharp tools produce good results. Glues well but very difficult to nail or screw. Takes stains but does not polish well. Natural finish with clear varnish or lacquer is better.	Handles for tools, wooden tools such as mallets, bench hooks, shooting boards, gauges and planes. Turnery and chair making. Wooden toys. Plywood.	Grain hard, close and very smooth—small medullary rays. Heavy wood. Pale brown colour. Has a tendency to warp and twist. Can be steam bent. Not durable out of doors.
English Elm	Great Britain.	Twisty grain may make working difficult. Sharp tools required to give a good finish. Nails, screws and glues well. Resists splitting. Readily accepts stains and polishes.	Furniture and cabinet work— coffins. Chair and stool seats. Turnery. Dock and harbour work, but not durable at ground level where both air and moisture are present.	Grain often very twisted and 'wild'. Medium–heavy weight. Heartwood light to warm reddish brown with a purple tinge at times. Sapwood cream or pale yellow. Rot resisting and very durable in favourable situations. Excellent burrs for decorative work.

A number of hardwoods are imported. The most common ones are: Japanese Oak, Japanese Elm, African Afrormosia, Agba, Emeri, (Idigbo), Iroko, Obeche, Utile, Walnut, Brazilian Mahogany, Imbuya, Freizo, Malayan Meranti, Jelutong, Indian and Burma Teak and Ramin from Sarawak.

QUESTIONS

1. (a) Compare the uses of ash and beech. (b) Why are they distinctly different?

2. (a) Why should brass screws and fittings be used on oak? (b) What should be used as an alternative to nails, in good quality work in oak? (c) Make a simple sketch showing an example of where this would be done.

3. (a) Describe the differences between the grain of ash, sycamore and elm.

4. (a) List the uses of beech. (b) Why is it particularly suitable for these purposes?

5. (a) List the main uses of the five hardwoods featured in the table. (b) Beside each one add the names of two specific items for which you would use that timber.

REVISION QUESTIONS

Tree growth

1. The growth of trees in the Northern Hemisphere is seasonal. (**a**) Explain what this means and how this affects the growth of trees. (**b**) When is the best time to fell a tree for timber, and why?

2. What effect do trees, or the lack of them have on the environment?

Conversion, seasoning and defects

3. Show four different ways in which a log may be quarter sawn.

4. What are the advantages and disadvantages of (**a**) natural seasoning (**b**) kiln seasoning?

5. What natural defects must be avoided when selecting a piece of timber? Draw up a list of these.

6. (**a**) What are 'timber shakes' and 'honey combing'? (**b**) Discuss the possible causes of each.

7. Discuss the probable effects of using unseasoned or badly seasoned wood for interior cabinet work or joinery.

Decay and insect attack

8. Find out how you would eradicate furniture beetles which have infested the plywood back and legs of a display cabinet.

9. You have identified dry rot in the corner of a floor in your home. It has caused decay in approximately 1 sq. m. of floor boards and to the joists below. Find out from a manufacturer's leaflet how you can eradicate this problem.

Softwoods and hardwoods

10. List the standard metric units of sale for softwoods and hardwoods.

11. Name five softwood trees and indicate the parts of the world where they grow.

12. Describe the ways in which softwoods can be made more durable for outdoor use.

13. Very sharp tools are required to work many softwoods satisfactorily. (**a**) Why is this so? (**b**) List the tools that would be used to make a nail box with lapped corner joints and cross halved partitions fitted into stopped housings. (**c**) Discuss the probable effects on the work if the tools are blunt.

14. (**a**) Design a piece of modern furniture to be made entirely of Red Deal. (**b**) Explain why this timber is suitable for such a design.

15. To make a garden gate which softwood would you choose? Give reasons for your answer.

16. Yew is a decorative softwood. List four pieces of work for which it would be suitable and make sketches of them. Add coloured shading to your drawings to indicate the grain.

17. (**a**) Describe the appearance of the English Oak tree. (**b**) Describe the timber obtained from it and say what it is used for.

18. (**a**) Choose two hardwoods and draw designs for different occasional tables to be made from each one. (**b**) Say why each design is so suitable for each timber.

19. (**a**) Design a cheeseboard to hold a tile for cutting on. (**b**) Which hardwood would you choose and why?

20. (**a**) Distinguish clearly between softwoods and hardwoods. (**b**) How do the trees and the timber they produce differ?

THREE

Veneers and manufactured boards

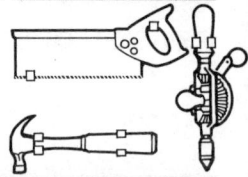

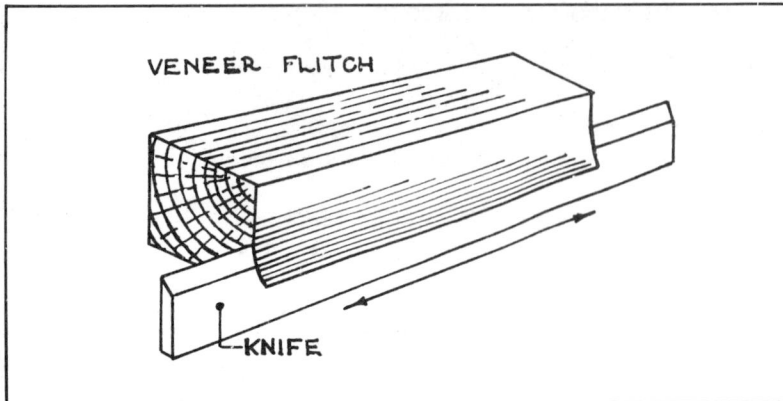

Fig. 148   Knife-cut veneers

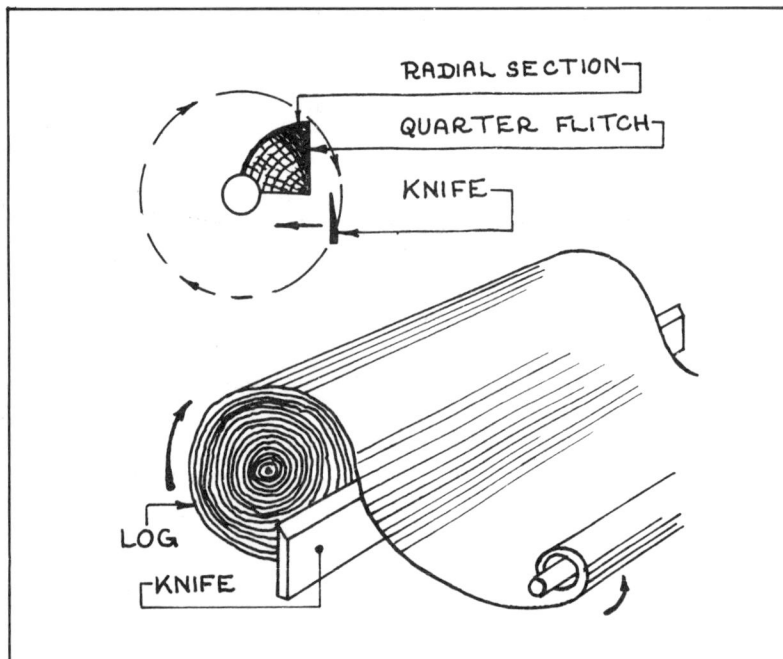

Fig. 149   Rotary slicing

3.1 VENEERS

Veneers are thin sheets of wood used in the manufacture of man-made boards, or to enhance the surface of cabinet work, e.g. boxes, cabinets, cupboards etc. Decorative veneers are cut from selected timber to show the full beauty of the grain.

Veneers are produced by one of three methods: sawing, flat knife cutting or rotary slicing.

Saw-cut veneers

These are very thick, extremely wasteful (almost 50 per cent wastage in sawdust) and so very expensive. They are now seldom used for decorative purposes.

Knife-cut veneers

The 'flitch' (Fig. 148) is the selected portion of the log from which veneers with the best grain formation can be cut. The 'flitch' is mounted on a flat bed. Then either by sideways movement of the bed, or the knife, veneers are sliced off. The knife is pitched so that it cuts across the 'flitch' at a slight angle to the grain. Many of the best veneers are cut in this way. The flitch is first steamed to assist cutting unless water stains would seriously affect the colour of the veneer.

Rotary slicing

If the veneers being cut are from a timber such as oak with pronounced medullary rays, and the veneer has to show well-figured grain they can be cut from a radial section of the 'flitch' (Fig. 149). The log is quartered and mounted at one of its outer corners so that as the knife moves up to it veneers are cut off approximately in line with the medullary rays.

Long continuous sheets of veneer are produced by rotating the whole log against the knife. This results in the veneer having a 'closed' side and an 'open' side when it is laid flat, so whenever possible the open grain should be glued downwards. The logs are about 2 m long. This is also the method used to cut veneers for plywoods.

QUESTIONS

1. **(a)** What is a veneer? **(b)** Why are veneers used in preference to solid timber?

2. **(a)** How are veneers cut from the log? Illustrate with simple diagrams. **(b)** Which method produces the largest sheet of veneer?

3. Name the parts of the tree from which the most decorative veneers may be cut. Add any necessary notes or sketches.

4. Make a list of those trees from which veneers are often cut.

5. **(a)** Why are veneers cut to different thicknesses? **(b)** How would you store veneers which are in stock?

3.2 HOW TO APPLY VENEERS

The part of the workpiece being veneered is called the '**ground**' and may be solid timber, plywood, laminboard, blockboard or chipboard. The 'ground' must be clean and free from irregularities that may be 'telegraphed' through the very thin veneer e.g. deep bruises, large broken knots, pronounced grain, nail holes, screw heads, or any joints used to hold the grounds together which show end grain, would eventually cause irregularities to the surface of the veneer. Care must be taken to ensure the ground is free from these and other defects. The ground must then be keyed for the glue either with the **toothing plane** (Fig. 150) or coarse hacksaw blade. Note the steep pitch of the blade so it will scrape.

Other tools required for veneering are few and simple (Fig. 151). The **veneer 'hammer'** is made of beech wood with a head approximately 100 mm wide into which a brass blade is set. Brass is used because it is non-corrosive and will not stain the veneers. A **trimming knife** with thin interchangeable blades is needed to cut veneers to size, to cut joints in the veneers and to trim off surplus, usually with the aid of a straight edge. An **electric iron** can be used to remelt the glue if necessary during the laying-on of the veneer, but should be disconnected once hot to avoid scorching.

To lay a veneer with a hammer (Fig. 152) both ground and veneer have to be sized with hot, thin Scotch glue. When dry, thin glue at 65° C is applied to each, the two are pressed together and the surplus glue squeezed out with the hammer. Light ironing at this stage ensures firm adhesion, and if necessary, by dampening the surface of the veneer, the glue can be remelted with the hot iron.

When large areas are to be veneered and sheets have to be edge jointed, the veneers should be overlapped and both thicknesses cut through with the knife (Fig. 153). The waste can then be removed with care, the edges pressed together and adhesive tape used to reduce the risk of shrinkage at the joint. Modern impact adhesives can be used quite successfully but are not suitable for the above technique. Cut the joint first, tape together and apply the whole piece as one.

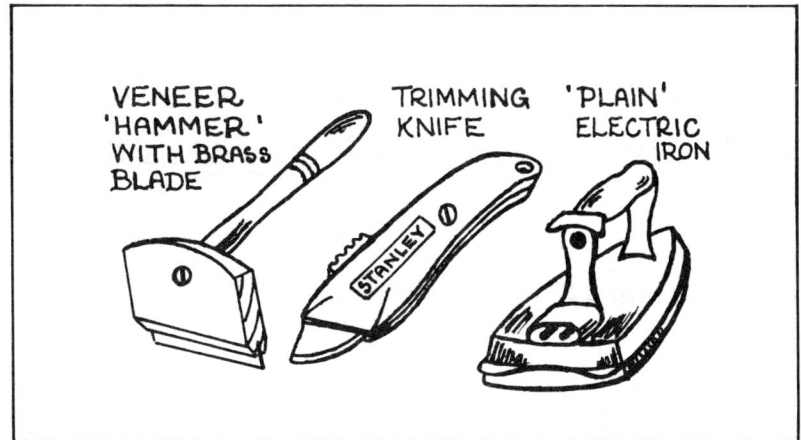

Fig. 151

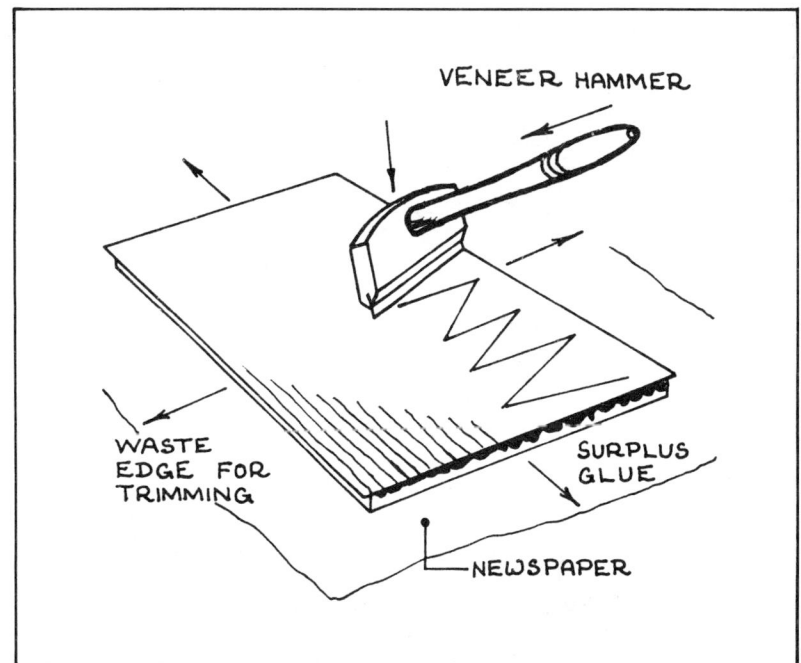

Fig. 152   Laying a veneer with a hammer

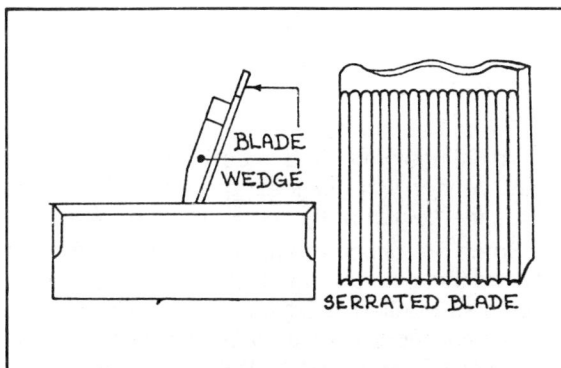

Fig. 150   Wooden toothing plane

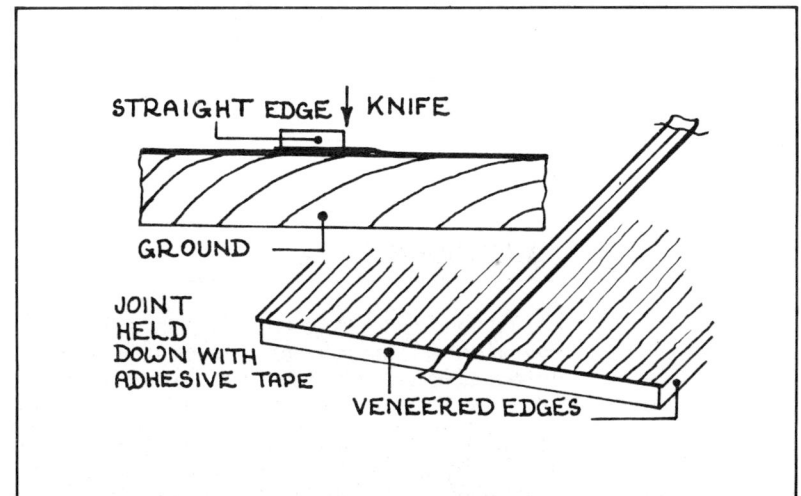

Fig. 153   Edge jointing veneers

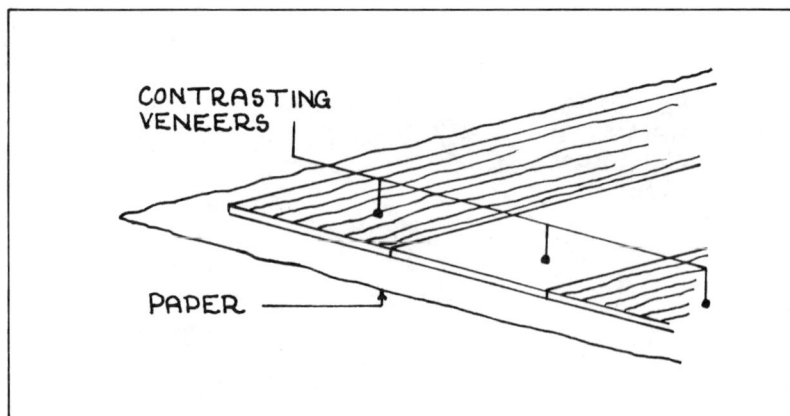

Fig. 154   Laying contrasting veneers

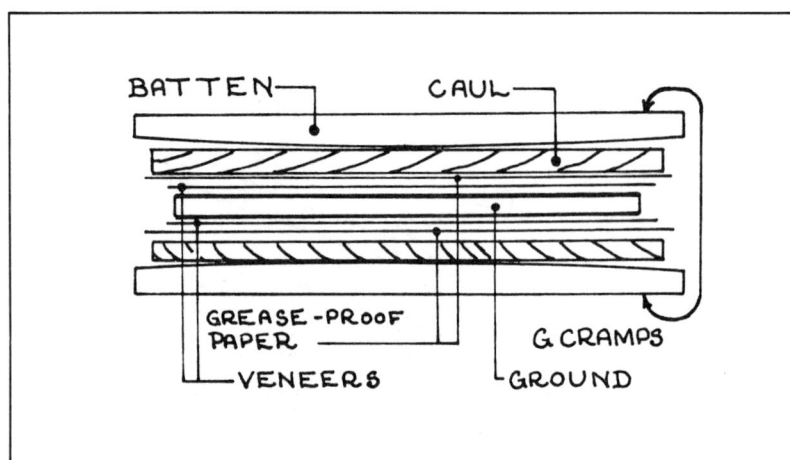

Fig. 155   Veneering with cau

If several pieces of veneer have to be laid to form a pattern or design (Fig. 154), it is good practice to glue them inverted onto a piece of stout paper under light pressure. The assembled veneers can then be glued onto the ground and the paper soaked off.

Veneering with PVA and modern synthetic resin glues can be done in a veneer press. Alternatively the work can be cramped between **cauls** (Fig. 155) (stout pieces of wood) pressure being applied from the centre outwards by curved battens. Grease-proof paper is used to keep surplus glue off the cauls.

QUESTIONS

1. (a) Veneers are glued onto specially prepared wood. What is the technical term for this? (b) What precautions should be taken with the preparation of this wood?

2. (a) How is the surface of the wood prepared for veneering? (b) Why is this necessary and what tools are used?

3. (a) Make a 3-dimensional sketch of a veneer hammer. (b) Why should it have a brass blade?

4. Write step by step notes to explain how you would veneer the lid of a small trinket box 150 × 150 mm square.

5. Make a sectional sketch to show how the edge-to-edge joint is obtained between veneers when large sheets are being laid.

3.3 MANUFACTURED BOARDS

The different types of manufactured boards are well known: plywood, hardboard, chipboard or particle board, and plastic laminates that are often referred to by their trade names. These man-made boards have many advantages, but some disadvantages.

Advantages

1. They are available in large sheets, and so big areas can be covered quickly. Also they can be used quite readily for large sections of a work-piece e.g. a table top.

2. They are economical—the whole board can be used, and although they are not cheap, the natural timber required to do the same work would be very much more expensive.

3. They are stable, resisting the tendency to twist and warp caused by changes in temperature and humidity.

4. They are available in many thicknesses.

5. They are made in a wide range of standard sheet sizes.

6. They are suitable for constructional purposes as large areas can be covered with few joints.

7. Also when used for constructional purposes most have the same or greater insulating and heat-resisting properties as the timber from which they are made. Some plywoods are treated chemically to make them fire-retardant.

8. Different specialist surface finishes are applied by the manufacturer and the most appropriate can be chosen, e.g. natural veneers, plastic wood-grain finishes moulded and textured surfaces, many of which are heat and moisture resistant.

Disadvantages

1. The most serious is that the edge of the board is unsightly and must be covered with paint, veneer, wooden lipping or a frame. In some cases the design of the work will hide the edge.

2. Some are difficult to fix e.g., it is very difficult to drive thin pins through plywood, and in thicker sheets pre-drilled holes are necessary. On the other hand fibre insulating boards are very soft and only nails with big heads will hold it which can be very unsightly.

3. Screws will hold in plywood, but should not be used on the edge, because the end grain of alternate laminations, could split open.

4. Chipboard will not hold standard screws very well. Only Twinfast screws should be used, but not on the edge.

5. 'Finishing' may be difficult in some cases and where possible 'pre-finished' boards should be used.

3.4 PLYWOODS

Plywood is always made up of an odd number of plies of wood veneers, so on the outer faces the grain runs the same way. In standard plywoods the grain of alternate layers is at right angles making the board rigid, stable and very strong. Cross grained in this way it overcomes the inherent weakness of natural timber to split along the grain.

Plywood never has less than three veneers and there may be as many as nineteen (Fig. 156). Those boards with more than three veneers are '**multi-ply**'. Plywoods with the middle veneer much thicker than the outer ones are called '**stout-heart**'.

Veneers for plywood are cut in continuous lengths from the log (see 3.1). For this process straight, large diameter logs are used. To assist the cutting process they are first softened by steaming or soaking in boiling water. After peeling, the veneers are clipped to length, dried, selected for quality and bonded together under heated pressure. Bonding is usually done with a modern synthetic resin glue. These glues will determine the suitability of plywood for particular uses and are identified by the code letters shown below.

Grades of plywood

Int. interior use
U.F. exterior use
M.R. moisture resistant
B.R. boil resistant
W.B.P. weather and boil proof (Fig. 157)

Plywoods can be bent within limits and simple joints such as laps and housings are possible. If the edge is planed, care is needed to avoid splintering the corner. The best method is to plane from each end to the centre.

Uses in the workshop

Plywood is used for panels, cupboard backs and bottoms, box and drawer bottoms, dust panels between drawers, trays, table tops, and flat profile work. Plywood is a reliable material for high-efficiency structures such as this box-section assembly for the body of a high performance sports car (Fig. 158).

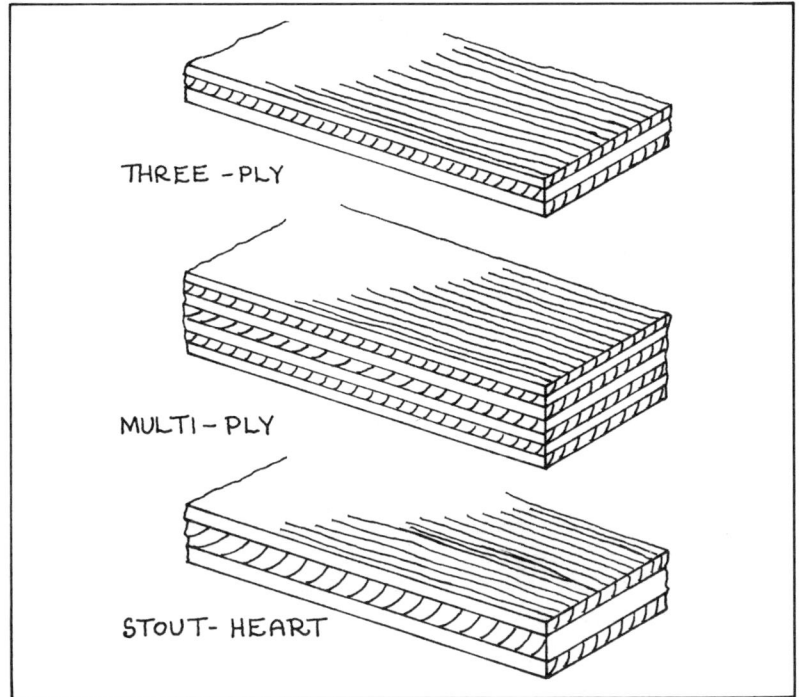

Fig. 156 Plywoods

Fig. 157 Boat building with W.B.P. plywood

Fig. 158

QUESTIONS

1. How is plywood made? Illustrate with sketches.

2. State two of the main advantages of plywood.

3. Explain how veneers are used in the manufacture of plywood.

4. How are veneers for plywood obtained?

5. Describe the workshop uses of plywood.

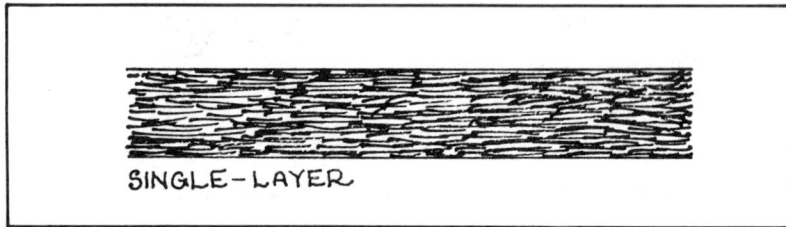

SINGLE-LAYER

Fig. 159

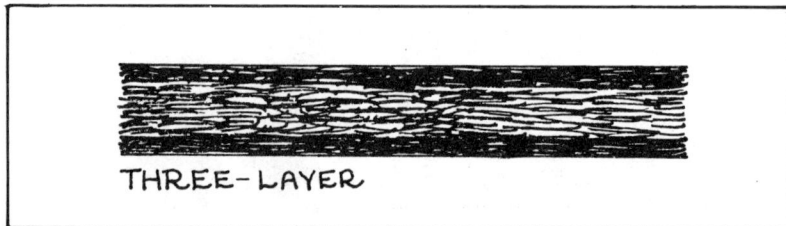

THREE-LAYER

Fig. 160

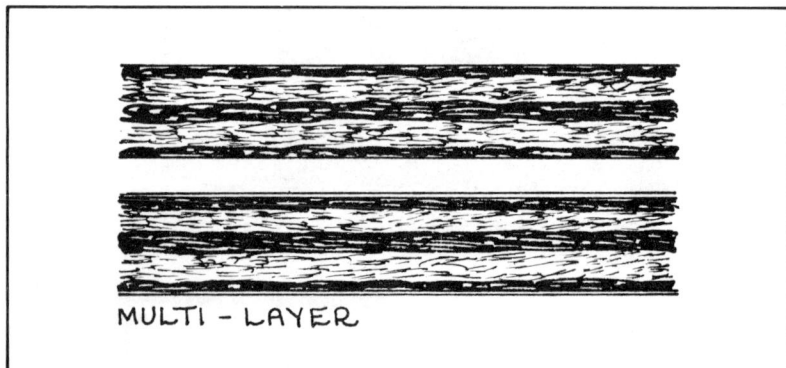

MULTI-LAYER

Fig. 161

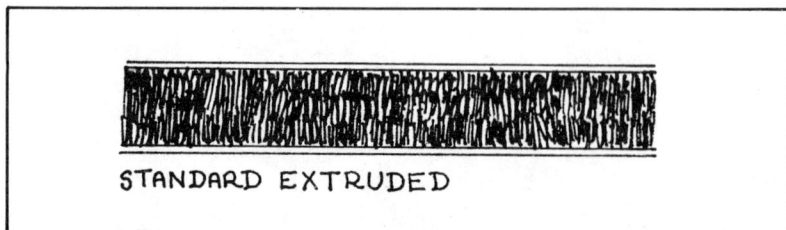

STANDARD EXTRUDED

Fig. 162

3.5 CHIPBOARDS

Ninety percent of chipboard is made from chippings obtained from spruce, larch, fir and pine, all softwoods. Chippings consist of forest thinnings and timber waste such as remnants from veneer making, offcuts, edge rippings, shavings and chippings from other timber users. These are reduced to chips of a required size by 'disintegrators' or 'coarse chippers'.

These wood chips are bonded together under pressure and heat, with a binder. Binders are synthetic resins or organic adhesives used to coat the wood chips before pressing or extruding them into a board. The density and structure of the board depends upon the type of woodchips used, the method of spreading and the pressure applied.

Any material introduced during manufacture to give the board special properties is called an additive. Preservatives, water repellents and fire retardants are in this category.

An outstanding characteristic of chipboard is its '**non-directional**' grain. Therefore pieces can be cut from any part of the board. It also has excellent gluing properties because the wood chips lie in a random pattern and there is no end grain. A good clean saw cut provides a consistently good gluing surface, regardless of the direction or angle at which it is cut.

Single-layer chipboard

This board (Fig. 159) has a consistent density being made from chips of the same size or same mixture of sizes. Its working properties are constant.

Three-layer chipboard

The outer surfaces are made from higher density, fine, long chips or flakes (Fig. 160). The core has larger chips and is therefore less dense. Being able to vary the size of the chips on the surface enables manufacturers to produce boards better suited to meet specific requirements e.g. a fine surface on which to paint.

Multi-layer chipboard

The strength may be further improved by introducing a central core of high density small chips, which also improves its flexibility (Fig. 161). The board may have five, seven or more layers, with layers of high density finer chips on both outer surfaces, giving a still better finish for paint or veneer.

Extruded chipboard

Using this method, boards of almost unlimited length can be produced with little or no waste (Fig. 162). Thicker boards than those produced by other methods can also be made.

In addition to chipboards with superfine surfaces for painting, there are boards with protective decorative surfaces of plastic laminates, plastic 'wood grain' type finishes and true wood veneers. Although veneers can be bonded to plain good quality chipboard in the workshop the best results are obtained by factory pressing. A urea-formaldehyde resin adhesive is recommended rather than PVA or traditional animal glue because their high water content can cause swelling of the surface particles, affecting the veneer.

Uses in the workshop

Plain chipboard is used for some carcase work such as shelving and other hidden parts, for occasional jigs and holding devices, and large constructions of a temporary nature such as stage sets. 'Veneered' chipboard has many uses such as table tops, doors and general carcase work for bookcases, cabinets etc. (Fig. 163).

Fig. 163   Wall unit made entirely from chipboard

QUESTIONS

1. (**a**) What is chipboard made from? (**b**) How is it manufactured?

2. What are 'binders' and 'additives'?

3. Distinguish between single and three-layer chipboards.

4. Chipboard is marketed with various surface finishes. What are they?

5. What are the workshop uses of chipboard? List as many examples as you can for each type.

3.6 LAMINATED BOARDS

Blockboard

This (Fig. 164) consists of a core, often made of redwood strips up to 25 mm wide, which are glued together to form a slab and sandwiched between two outer veneers. There are several variations of this construction, one of which is shown (Fig. 165).

Laminboard

The inner slab (Fig. 166) is made of much smaller strips of only 1.5 to 7 mm thickness, which results in a heavier board because of the extra glue required. In many cases denser timber is used for the core. It is therefore more expensive, restricting its use to high-quality work. Unlike blockboard where any irregularities in the core may be 'telegraphed' through the face, laminboard overcomes this because such small sections are used for the core.

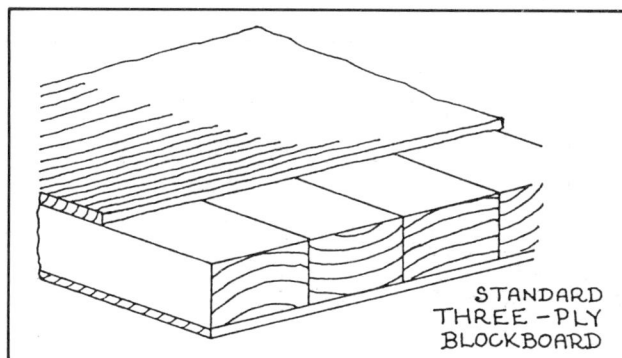

STANDARD THREE-PLY BLOCKBOARD

Fig. 164

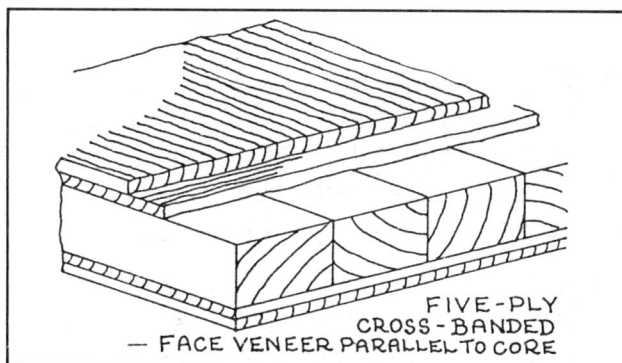

FIVE-PLY CROSS-BANDED — FACE VENEER PARALLEL TO CORE

Fig. 165

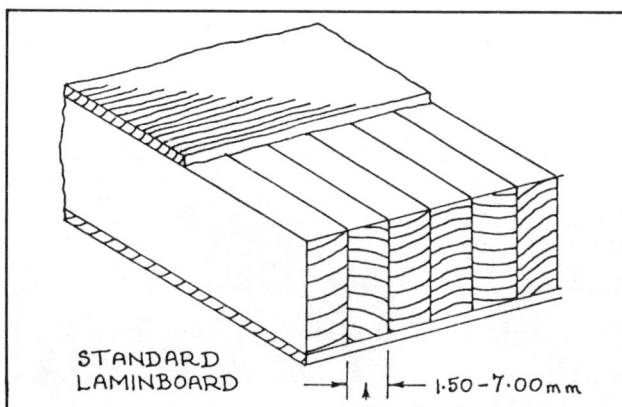

STANDARD LAMINBOARD ⊢ 1.50 – 7.00 mm

Fig. 166

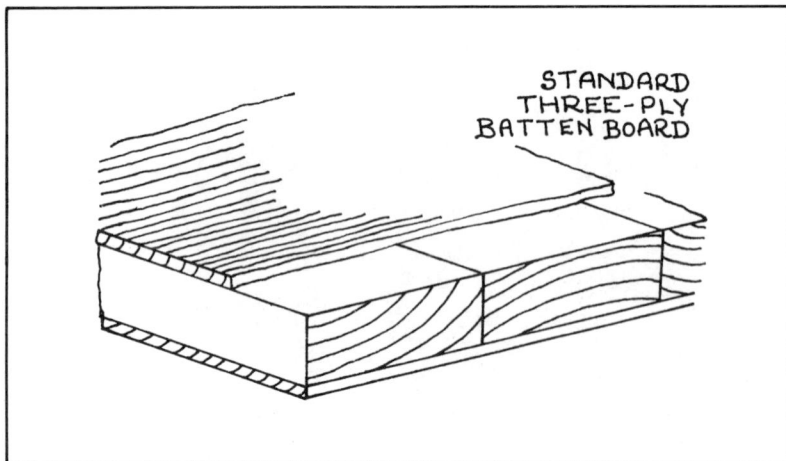

Fig. 167

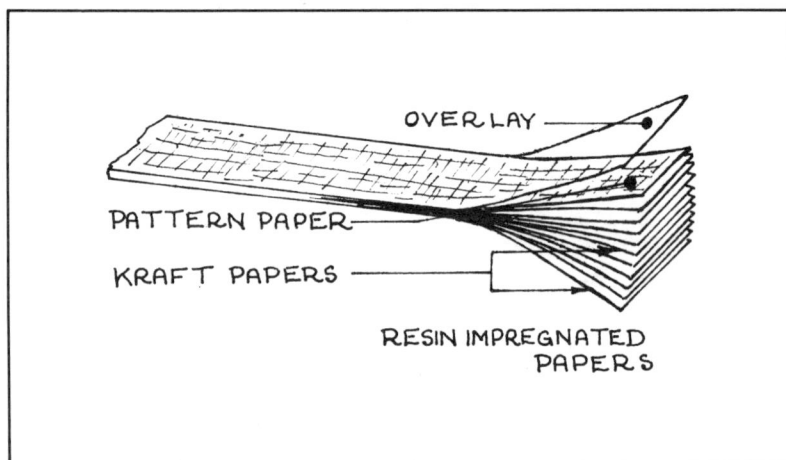

Fig. 168   Plastic laminate

Fig. 169   Wood-grain paper being impregnanted with resin

Battenboard

This is a cheaper variety (Fig. 167), the core being made of strips up to 75 mm wide which have an even greater tendency to show through the veneer.

Low density timbers such as pine or obeche reduce the weight of laminated boards if used for the core.

Uses in the workshop

Suitably selected they are ideal for slab constructed furniture, table tops, desk tops, cabinet doors and other constructions where rigid panels are required.

QUESTIONS

1. Name the different laminated boards and draw sectional sketches to show the differences between them.

2. Why is laminboard the most expensive type of laminated board?

3. What does 'telegraphed' mean in relation to veneers and laminated boards?

4. List the workshop uses of laminated boards. Include examples not given in the book.

3.7 HEAT-RESISTING LAMINATED PLASTICS

Modern high-pressure thermosetting laminates are so called because they are built up of resin impregnated papers which fuse together when subjected to the combined effect of high pressure and heat. Hydraulic presses are employed in much the same way as for hardboard, plywood, chipboard and blockboard.

Laminates (Fig. 168) consist of several layers of '**kraft**' **paper**, a **patterned** or **coloured paper**, and an alfacellulose paper or '**overlay**'. These are impregnated with phenolic or melamine resins (Fig. 169), oven dried and the required number 'laid-up'. This number depends upon the thickness of the laminate being produced. '**Glassine**' paper which does not absorb resin is used to separate the laid-up 'books' of paper during pressing.

Uses in the workshop

Laminates can be used for table mats, cheeseboards, small work tops and table tops where a heat and stain resisting surface is required. Bond with impact adhesive. (See 4.3)

QUESTIONS

1. How are heat-resisting laminated plastics made?

2. (a) What are laminated plastics made from? **(b)** How are they 'laid-up'?

3. List as many examples as you can where laminated plastic sheet would provide a suitable working surface.

3.8 HARDBOARDS

These are fibre boards manufactured from wood or woody plants. The fibres are 'felted' together with adhesives and water to produce a pulp which is then subjected to great pressure. The characteristics of hardboard produced in this way are density, smoothness and strength. Hardboard can be worked as a timber but tools need to be very sharp, and only fine toothed saws should be used.

Uses in the workshop

Hardboard is suitable for table mats, light panel work for painted cabinets and doors, racks and containers.

QUESTIONS

1. How is hardboard made, and what are its characteristics?

2. Hardboard can be worked in similar ways to other manufactured boards, within certain limits. What are these limits and give reasons for these.

3. (a) What are the workshop uses of hardboards? **(b)** For what other work would it be suitable?

REVISION QUESTIONS

1. (a) Design a piece of work to be veneered. Make good clear sketches leading to a 3-dimensional sketch from which to work. **(b)** Give step by step details of how the veneering would be carried out.

2. In making the playing area of a chessboard from contrasting veneers, how would you prepare and glue down the pieces of veneer?

3. (a) What are cauls? **(b)** Design a piece of curved work to be built up in this way, from several layers of veneer.

4. What are the advantages of manufactured boards?

5. (a) What are the disadvantages of manufactured boards? **(b)** How would you overcome these?

6. What advantages does laminated board have over chipboard? Explain clearly using sketches to illustrate your answer.

7. (a) How would you bond a laminated plastic sheet to a small table top made from blockboard? **(b)** At what stage would you cover the edges and why?

8. Design a small 3-dimensional piece of work to be made entirely from hardboard. Different thicknesses may be used, or be obtained by gluing pieces together. What 'finish' would you use?

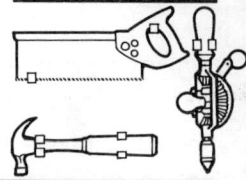

FOUR

Assembling and finishing

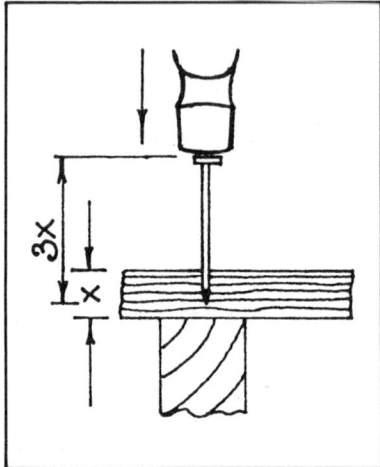

Fig. 170

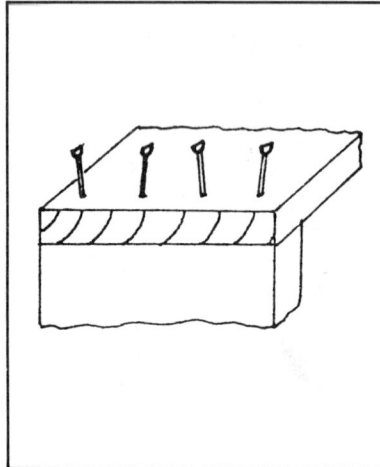

Fig. 171

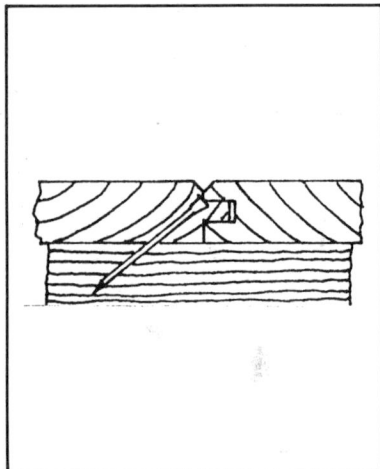

Fig. 172

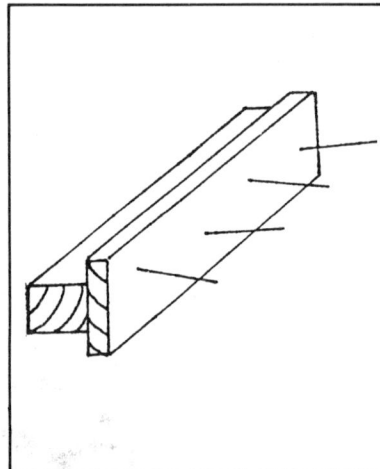

Fig. 173

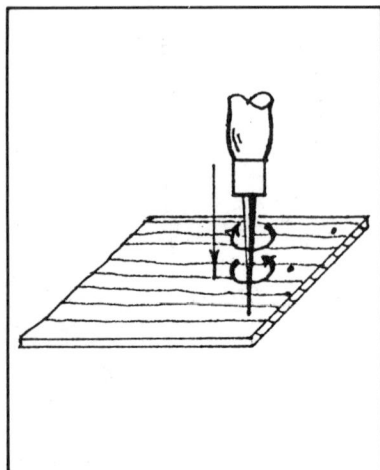

Fig. 174

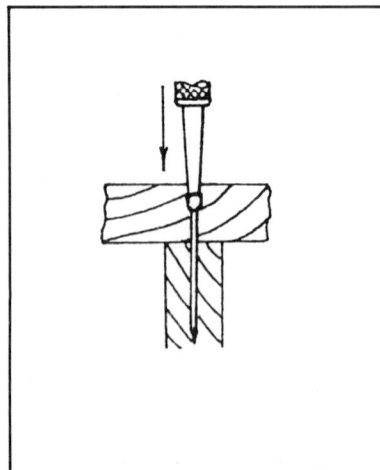

Fig. 175

4.1 NAILS

Nails fasten wood together quickly. They were originally hand-forged from iron; later they were cut from sheets of rolled iron. Mass production of nails began in the nineteenth century with the introduction of automatic machines producing nails from coils of wire.

Finishes

Depending on the use of the nail there are a number of finishes.

1. To remove grease and wire 'nippings' (particles left on the nail by the forming process) nails are cleaned in a rotating barrel of hot caustic soda.

2. Tumbling in a second drum of hot sawdust gives nails a bright finish.

3. Sherardizing will make them corrosion-resistant. A thin coating of zinc is applied by heating the nails to about 300° C in a closed container of zinc dust and zinc oxide.

4. A more permanent method of rust-proofing is to galvanise the nails in a bath of molten zinc.

Using nails

Care is needed when using nails and there are a number of points to remember.

1. The nail should be at least three times as long as the thickness of the wood it has to hold (Fig. 170). Always nail the thinner wood to the thicker piece.

2. Always hit the nail squarely and keep the hammer head clean to avoid foul blows.

3. Slope the nails in a dovetail fashion to increase their holding power (Fig. 171).

4. Tongued and grooved cladding can be secret nailed to hide the heads (Fig. 172).

5. *Do not put a row of nails in line.* This may split the wood *along the grain.* Stagger the spacing, which will pull the two pieces more closely together (Fig. 173).

6. The risk of splitting is reduced if the nail point is slightly blunted with the hammer so it tears through the fibres. Alternatively it may be advisable to drill small holes with the hand drill or bradawl (Fig. 174).

7. Nails with small heads can be punched below the surface (Fig. 175). The hole can then be filled before the finish is applied (See 4.8).

Types of nail

Round wire nail flat round head; diamond point;

length 13 to 304 mm; carpentry and joinery, general construction work.

Oval brad available with 'lost' head; diamond point; length 13 to 150 mm; risk of splitting is reduced because of elliptical section; head is easily punched below the surface of the wood.

Lost head nail round wire floor brad; diamond point; length 38 to 100 mm; used for flooring because head is easily punched in, and for similar work to round wire nail where the flat head would be unacceptable.

Panel pin deep countersunk head; diamond point; length 13 to 75 mm; slender nail for cabinet work where the head can be easily punched in and concealed with a filler.

Hardboard pin diamond blob head and diamond point; length 13 to 38 mm; for fixing thin panels; head is driven in and 'filled'.

Gimp pin plain flat head and diamond point; length 13 to 25 mm; a very fine nail for very light construction work.

Escutcheon pin convex head and diamond point; length 13 to 38 mm; used to fix escutcheon plates over key holes on doors.

Upholstery nail large domed head and diamond point; length 13 mm; used to fix upholstery coverings—leather, velvets, fabrics etc. to the frame.

Cut tack flat head and long tapering point; 6 to 34 mm for fixing folds in upholstery where an ornamental head is not required.

Plaster board nail large diameter, thin flat head to support the board; length 25 to 75 mm; diamond point.

Ringed shank flat head and diamond point; length 19 to 200 mm; heavy duty fixing of sheet materials; ringed shank is gripped firmly by the fibres of the wood into which it is driven.

Corrugated fastener wriggle nail; various lengths and depths; for quick fixing of butt-jointed frames; care must be taken to avoid splitting with the grain; casual jointing of boards edge to edge (Fig. 176).

Pincers

They are used to pull out bent or badly placed nails (Fig. 177). Care must be taken to avoid bruising the surface of the wood. It is good practice to use some scrap wood both as protection and a fulcrum on which to lever.

QUESTIONS

1. Name and sketch the three most common types of nails. Your sketches should be large enough to clearly show the main details of each.

2. (a) Why do nails have different 'finishes'?
(b) Give details of what they are.

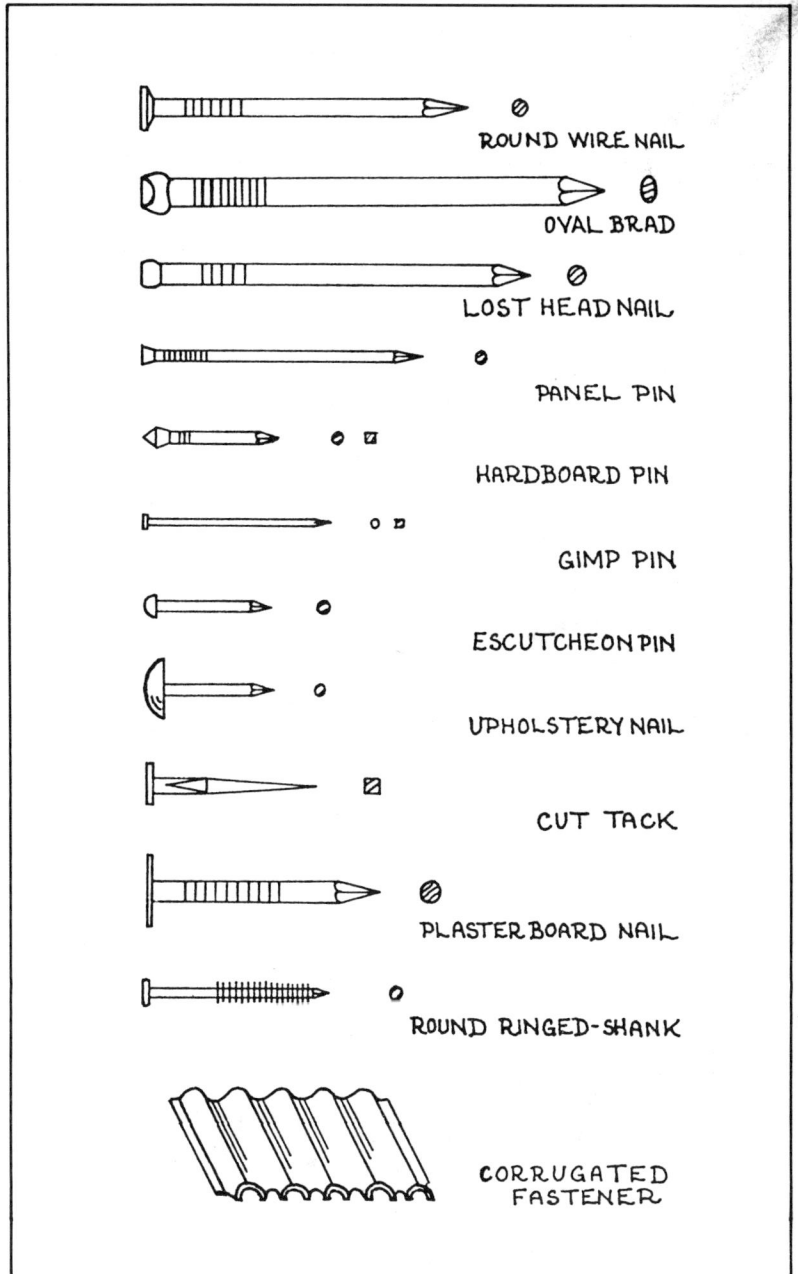
Fig. 176   Types of nail

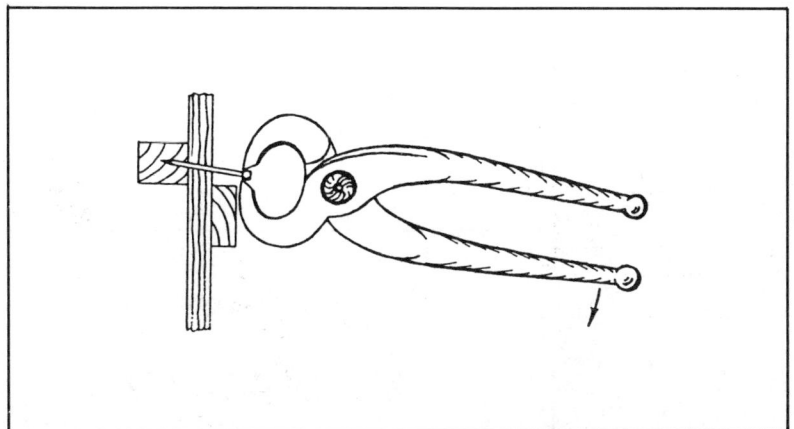
Fig. 177   Pincers

3. What points should you remember when using nails? Illustrate your answer with suitable sketches.

4. (a) What can be done to reduce the risk of splitting the wood when driving in nails?
(b) How are bent or unwanted nails removed?

5. (a) Name the nails with heads small enough to be punched below the surface of the wood. Why is this done? (b) What should be done with the resulting hole?

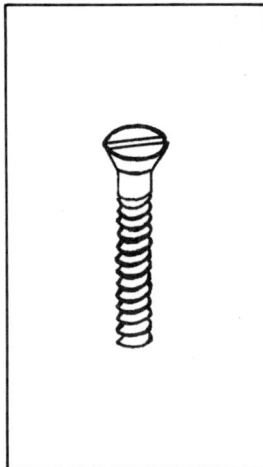

Fig. 178   Before 1854

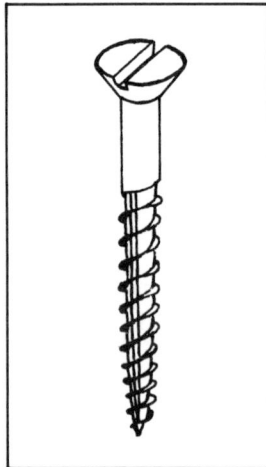

Fig. 179   Conventional gimlet point woodscrew

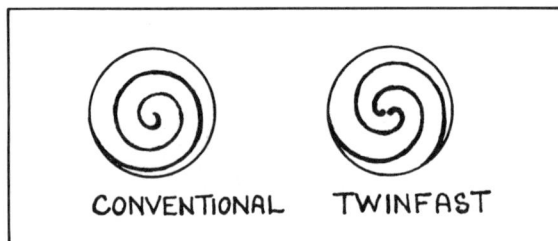

Fig. 180   Conventional and 'twinfast' threads

Fig. 181

Fig. 182

4.2 SCREWS

Because of their screw-thread, screws pull pieces of wood tightly together, the joint remaining firm and strong. Unlike nails they can be taken out comparatively easily if necessary, without any damage to the surrounding material. They will pull glued surfaces closely together making an even stronger and more permanent joint. As the wood-screw is driven into the side grain of the timber it forms its own mating thread with the fibres.

It is very bad practice to screw into the end grain. As the screw is finally tightened the threads sever the fibres and the screw loses its holding power. The threads are said to be stripped and the screw can be pulled out with little effort. A screw can be made secure in end grain by inserting fibre or plastic plugs into the end of the wood. Another way is for a dowel to be glued in across the member so that when the screws are inserted, they engage the side grain of the dowel and nip up tightly.

Screws are made from mild steel, brass, stainless steel and occasionally aluminium alloy. Mild steel, and brass screws may have protective and decorative finishes of chromium, zinc or cadmium, or be 'brassed' or 'blued'. Mild steel round headed screws are almost always 'black japanned'.

Types of woodscrews

Early screws had no gimlet point (Fig. 178). In 1854 Nettlefolds introduced the conventional woodscrew with a gimlet point (Fig. 179). This meant small gauges, particularly in softwoods did not need pilot holes because the gimlet point would penetrate the wood. These screws have good holding power in natural wood, but are not very effective in particle boards.

Conventional screw
This has a single start, gimlet point thread (Fig. 180). The screw is threaded for two thirds of its length only (Fig. 179). The shank may cause splitting because its diameter is equal to the outside diameter of the thread, unless an adequate clearance hole is drilled through the first piece of wood.

'Twinfast' screw
This is self-centring, with twin parallel threads (Fig. 180), and balanced penetration so it can be driven in accurately. The longer thread gives greater grip particularly when used to secure thin fittings such as hinges etc. These screws are available with CSK, raised CSK or round heads.

GKN 'Supafast' screw
This is recommended for all types of man-made boards and natural timbers. It is available as GKN 'Mastascrew' (Fig. 181) with slotted heads of traditional shapes, or GKN 'Supascrew' (Fig. 182) with recessed heads. Both types of 'Supafast' screw are available with CSK or round heads. These screws now supersede the previous

Twinfast varieties. Both types of 'Supafast' screws have the following characteristics.

1. The steeper pitch of the parallel twin threads enables the screw to penetrate the wood quickly and easily.

2. The point is sharp and self-centering, therefore the screw starts easily without slipping.

3. Along its length the shank diameter is less than the diameter of the thread so reducing the risk of splitting.

4. The whole screw is case-hardened. This gives added strength, and minimises damage to the head by the tip of the screwdriver blade.

5. The modified recess in the head provides better seating for the point of the Pozidriv screwdriver.

6. Each screw is corrosion resistant because of the standard electroplated finish.

Screw heads

The name of the screw is determined by the shape of its head (Fig. 183). The '**countersunk**' (CSK) head is used where a flush finish is required. The screw head fits into a predrilled countersunk clearance hole. It is good practice to countersink the underside of the clearance hole to take up any chips turned out by the screw.

'**Round**' headed screws are mainly used to fix thin metal fittings to wood, the flat underside of the head giving support to the metal.

'**Raised CSK**' headed screws are used if a more decorative and attractive appearance is desirable, and are usually of the slotted conventional type.

Screw cups

The surface fitting type (Fig. 184) gives support to thin material where countersinking may not be possible. Inset screw cups maintain a flush surface and prevent the screwhead from pulling in too far.

Screw cover caps and heads

They are designed to conceal screw heads (Fig. 185). Made of white or coloured plastic they may be a simple press fit into a counterbored hole, or a snap fit over a moulded washer. Cover heads screw into the screw head and are polished or satin finished brass, or chrome on brass.

Heavy-duty fixing

The **coach screw** (Fig. 186) is used where conventional woodscrews are neither big enough nor strong enough. They are very good for fixing heavy metal fittings to wood, but a washer is required under the head for wood to wood fixtures. A pilot hole should be drilled and counterbored for the shank. The screw is driven home with a

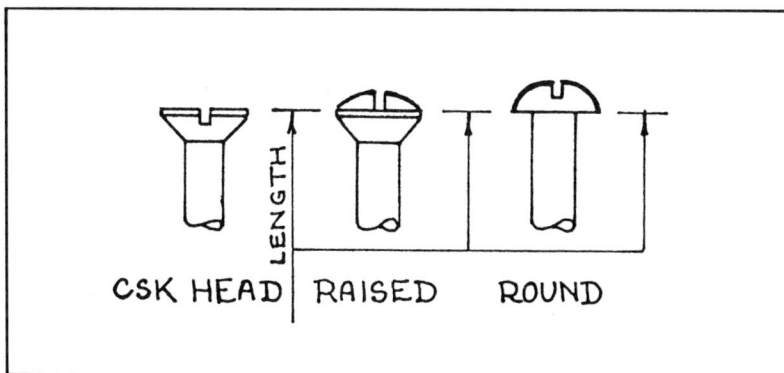

Fig. 183   Screw heads

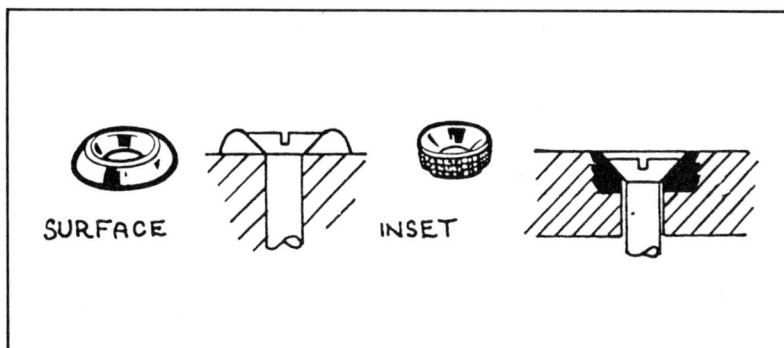

Fig. 184   Screw cups

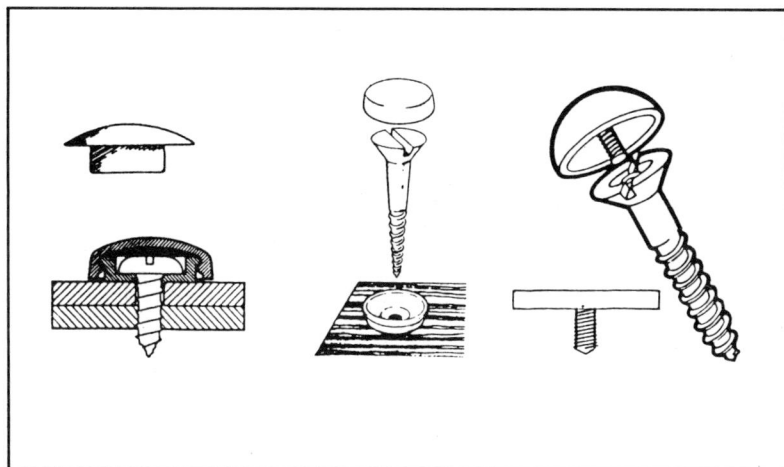

Fig. 185   Screw cover caps and heads

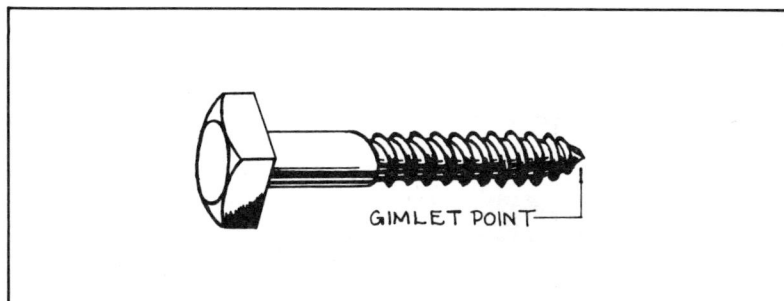

Fig. 186   Coach screw

spanner. They are available 'black' or 'galvanised' in lengths up to 300 mm.

A **carriage bolt** (Fig. 187) can be used as an alternative to the coach screw but a clear hole must be drilled through the work. The square section of the shank locks the bolt head in the wood, but a washer is required under the nut. Made in lengths from 20 to 400 mm.

Using screws

1. Choose the right screw for the work:
(**a**) The **length** when fixing wood to wood, should not be less than twice the thickness of the top piece with seven and no less than four threads screwed in (Fig. 188).
(**b**) The **diameter**, by screw gauge, should not be more than one tenth the width of the wood being fixed.

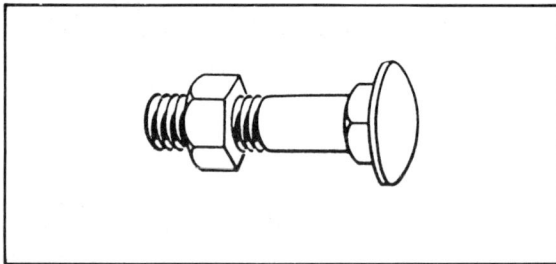

Fig. 187   Carriage bolt

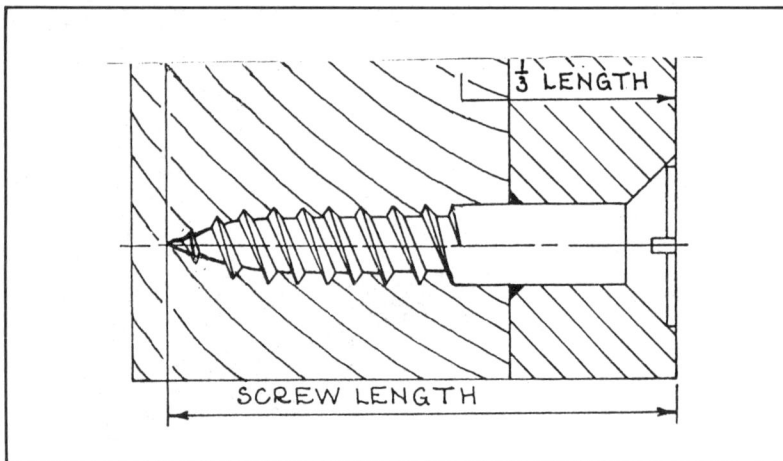

Fig. 188

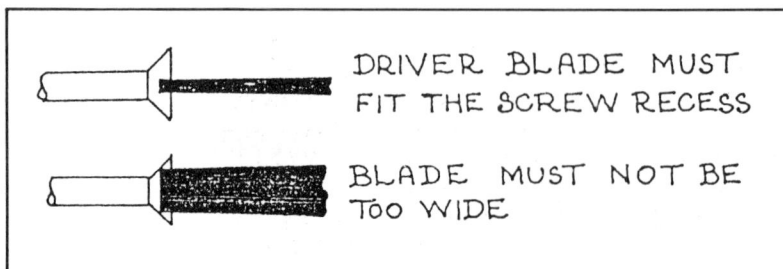

Fig. 189

(**c**) The **head** must be the most suitable shape for the work.
(**d**) The **metal** and '**finish**' of the screw must be suitable.

2. Use a good fitting screwdriver (Fig. 189); longer screwdrivers give more torque (turning power).

3. If brass screws are being used, first use steel ones of the same size to cut the thread. This applies particularly in hardwoods when the head of the brass screw could be damaged or worse still the whole screw could be 'twisted-off' in the wood.

4. Never use steel screws in oak because of the tannic acid it contains; purple and black stains soon develop. Use brass screws but note **3** above.

5. If the screw is difficult to turn, or worse still in danger of 'twisting-off', take out and lubricate with tallow or beeswax; again note **3** above.

QUESTIONS

---

1. Explain how the thread of the woodscrew pulls pieces of wood closely together.

2. (**a**) Why is it bad practice to screw into end grain?
(**b**) Explain how you could overcome this difficulty.

3. (**a**) Sketch the different shapes of screw head.
(**b**) When would you use the different screw heads? Illustrate with separate sketches.

4. Contrast the screw thread of the conventional screw with that of the modern 'twinfast'.

5. (**a**) Explain why, when screwing two pieces of wood together, a clearance hole is needed through the top piece. (**b**) When would you also make a pilot hole in the second piece? (**c**) What special precautions should you take when using brass screws?

4.3 ADHESIVES

Surfaces to be joined must be as flat and smooth as possible and, particularly in the case of joints, close fitting. There is some penetration of the wood by the glue, but the real strength of the glued joint comes from the closeness of the glue to the wood as a whole. There must always be sufficient glue to fill the joint to ensure good adhesion.

PVA—Polyvinyl acetate

This is probably the most widely used adhesive in the school workshop. This glue is a white emulsion rather like cream and is thermoplastic i.e. cold setting. These glues are only satisfactory for interior work where the atmosphere will be free from moisture.

Wood being bonded should be dry. The adhesive need only be applied to one surface except in the case of high density timbers such as oak and teak, when it should be applied to both to ensure 'wetting-out'. Oily woods should be de-greased with l.l.l. trichloroethane or methylated spirits.

The joints should be cramped up while the glue is still wet. *These adhesives set by loss of water* therefore the setting time depends on the absorbency, moisture content, temperature of the wood and the pressure which determines the thickness of the glue line. Many joints can be handled after as little as 10 to 40 minutes clamping but full strength is not reached for several hours.

All traces of surface glue must be wiped away from the joints with hot water as soon as the work is cramped up. Failure to do this will result in unattractive white glue stains if a clear finish or stain is applied to the work. Another way is to mask the work with sellotape, or to 'finish' each member before assembling.

Fig. 190 Laminating Workshop—main supports are laminated frames of Douglas Fir bonded with Aerodux

Urea-formaldehyde adhesives

These are 'gap-filling' glues made from synthetic resins and hardened by chemical reaction. The two-part pack is for separate application i.e. the adhesive is applied to one part of the work, the hardener to the other. The glue sets when the two parts are cramped together.

Aerolite 306

This adhesive comes in powder form and is mixed with water to a syrupy consistency. It has the advantage of a two year shelf life. At all times accidental contact between resin and hardener must be avoided, and the applicators kept separate which can be made from crushed cane and colour coded. Aerodux (Fig. 190) is an adhesive of this type. **Cascamite** is a convenient form of the same type; adhesive and hardener are combined in one powder to be mixed with water. Both are water resistant.

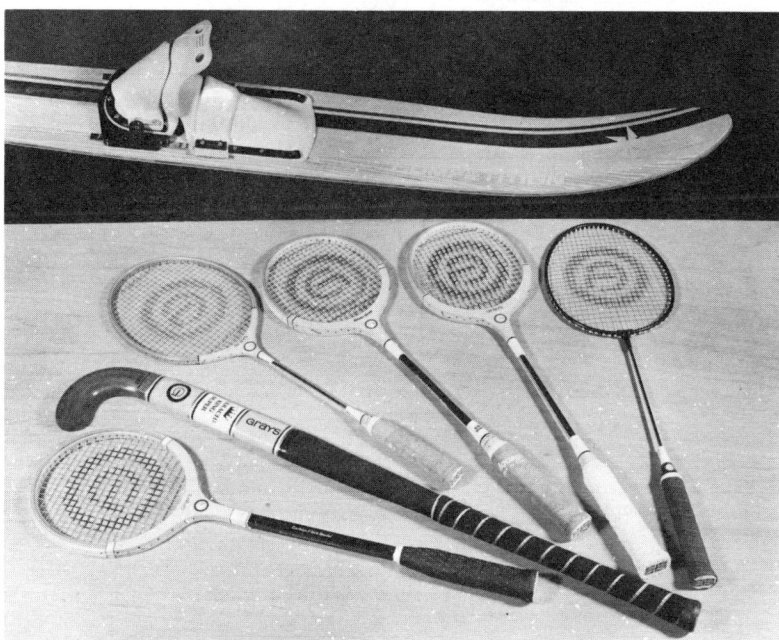

Fig. 191 Araldite used for sports goods

Epoxy resins

These are supplied as a twin-pack; equal quantities of resin and hardener are mixed together. Although there is an immediate chemical reaction the mixture has a usable life of about three hours. Maximum strength is developed in about three days. Epoxy resins will join many materials successfully, however the advantage to the woodworker is that bonds between wood and dissimilar materials, like metals and plastics, can be successfully made (Fig. 191). Items being joined must always be grease-free. Rapid setting epoxy resins are also available.

Impact adhesives

These are used in the school workshop to bond veneers and laminated plastics to flat 'grounds' of chipboard etc. Both surfaces must be covered evenly using a 'toothed spreader'. When the adhesive is 'touch-dry' i.e. it does not transfer if touched lightly with a knuckle, the parts are brought together. Adhesives which allow 'slip' for adjustment are preferable. Pressure should then be applied from the centre outwards to make the bond.

Scotch glue

This is now little used in the school workshop. Traditionally it is the glue used for veneering (See 3.2). Glue pearls are soaked, and melted with water in a glue pot; boiling water in the outer pot brings it to a working consistency.

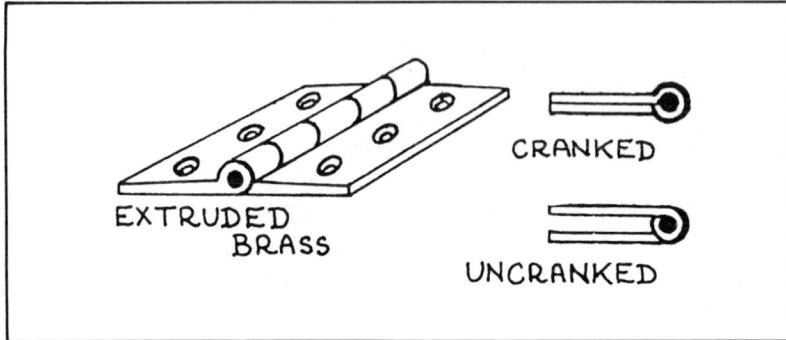

Fig. 192   Butt hinge

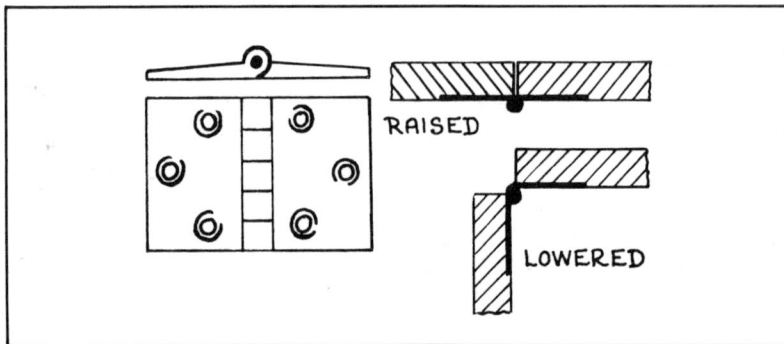

Fig. 193   Back flap

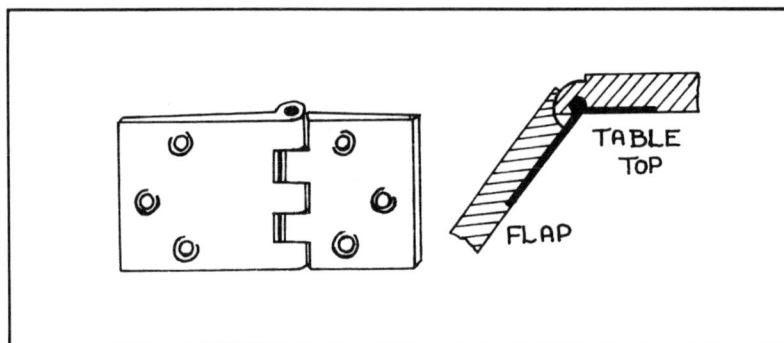

Fig. 194   Table hinge

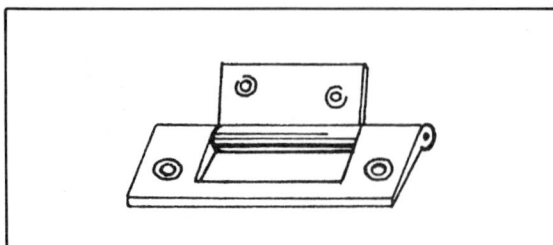

Fig. 195   Flush hinge

QUESTIONS

1. **(a)** What are the requirements for effective gluing. **(b)** What special precautions should be taken with high density timbers such as oak, and oily timbers such as teak?

2. **(a)** What is PVA? **(b)** How can you prevent it from making glue stains on your work when assembling?

3. As PVA adhesives set by loss of water, what things determine their setting time?

4. **(a)** Distinguish between urea-formaldehyde adhesives and epoxy resins. **(b)** What is the particular advantage of epoxy resins?

5. **(a)** What is an 'impact adhesive'? **(b)** When would you use an adhesive of this type, and how is it used?

4.4 HINGES

In earlier times a woodworker made his own metal fittings or they were made by the local blacksmith, as the need arose. As more items of furniture were made from wood, the need for a greater variety of fittings grew.

Originally these were made from iron, then mass produced from iron and steel. Brass was gradually introduced for quality fittings. Today increasing use is made of nylon and plastics. Fittings may have a protective plating of brass, chromium, zinc, copper or nickel, or have decorative finishes of antique brass, satin-brass, or bronze.

Butt hinge

Of all the many different types made, the butt hinge (Fig. 192) is the most common and most frequently used. The best quality butt hinges are made of 'extruded' brass. When fitted they must be fully recessed or 'boxed'. Steel butts may have cranked or uncranked knuckles which affect the 'boxing'. The cranked knuckle is usually left fully protruding, the uncranked only half.

Back flap

The back flap (Fig. 193) is similar to the butt hinge but is wider. It is used on bureau flaps, table leaves and deep lids where a moulding is not involved and is available in steel or brass.

Table hinge

This is specially designed (Fig. 194) for use on table drop leaves with a 'rule-joint' moulding.

Flush hinge

Made from plated steel or extruded brass (Fig. 195), it is easy to fit as it does not have to be 'boxed' in. It is strong with an easy movement and can be used instead of a butt hinge.

Tee hinge

This is used (Fig. 196) on rough boxes, chicken hut and rabbit hutch doors, light wooden gates etc. It is normally made from black japanned mild steel for outdoor use.

Piano hinge

Made from steel and bright electro brassed, or nickel plated, the piano hinge (Fig. 197) is widely used on modular units, and cabinets with slab doors. It is also available in plastic.

Concealed hinges

Lay-on concealed hinges (Fig. 198) are easy to fit, with the hinge screwed to the door and stile, and the slotted holes simplifying alignment. It is spring loaded to hold the door firmly open or securely closed without the need for catches or stays.

Concealed spring hinges (Fig. 199) are recessed into the back of the door and have an arm connected to the cabinet side by a separate base plate. They are used on 'lay-on' doors i.e. doors which fit in front of, and fully or partially conceal the cabinet frame. The base plate is screwed to the cabinet carcase and the arm of the hinge is connected to it. When the door is closed the hinge is totally concealed.

QUESTIONS

1. Using labelled sketches show the differences between a butt hinge and a backflap.

2. Draw sectional sketches across the knuckle to show how a butt hinge made from pressed steel differs from one made of extruded brass.

3. (a) Draw a sketch of a tee hinge. (b) Where would you use it and with what kind of screw would you fix it?

4. (a) Sketch a flush hinge to show how it differs from the butt hinge. (b) Why is it so easy to fit?

4.5 CATCHES, BOLTS, LOCKS, KNOBS AND HANDLES

Catches and bolts

These are used (Fig. 200) on small light-weight doors on cupboards and cabinets, to hold the

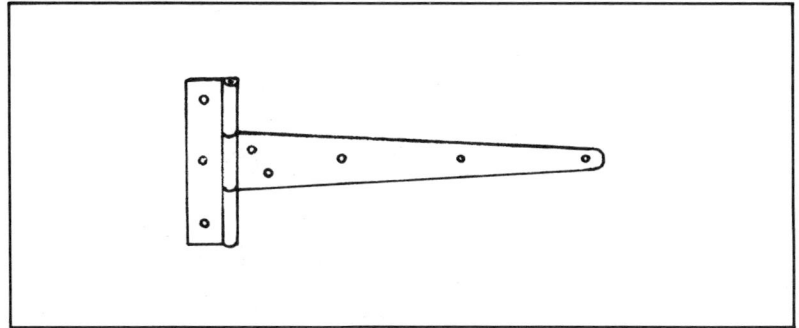
Fig. 196   Tee hinge

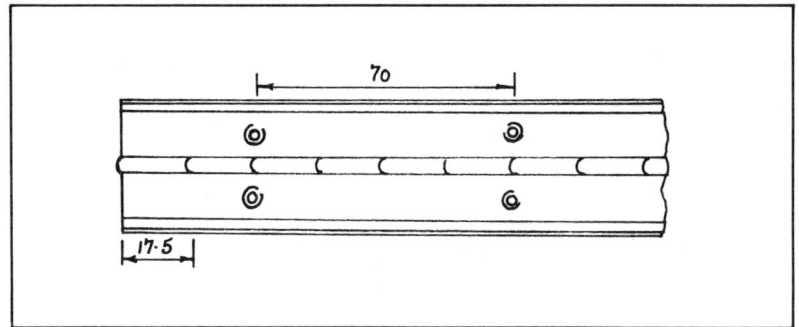
Fig. 197   Piano hinge

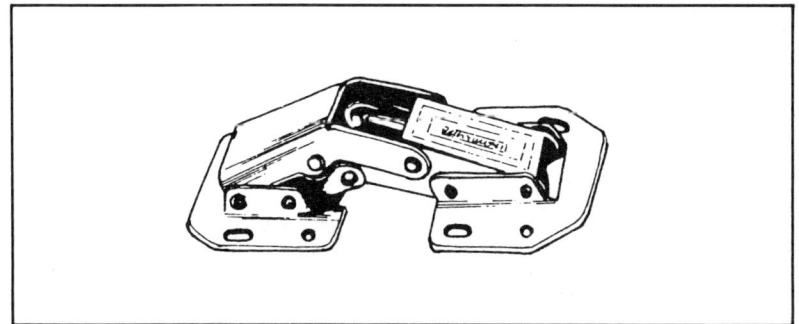
Fig. 198   Lay-on concealed hinge

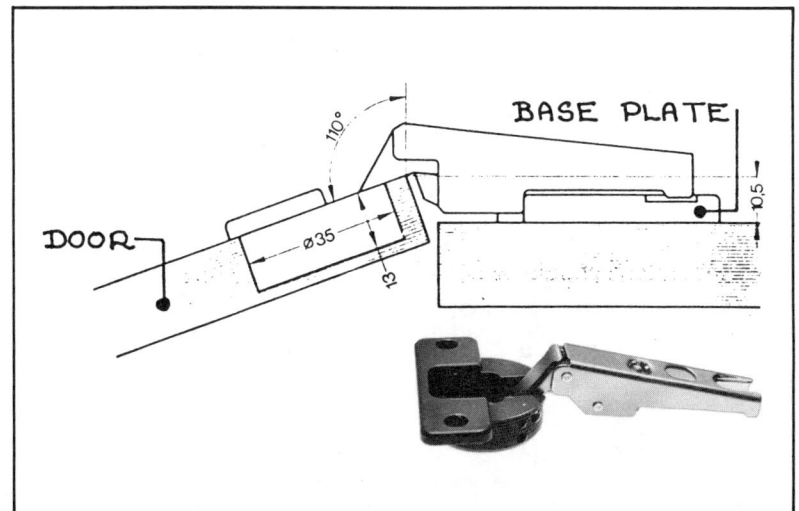
Fig. 199   Concealed spring hinge

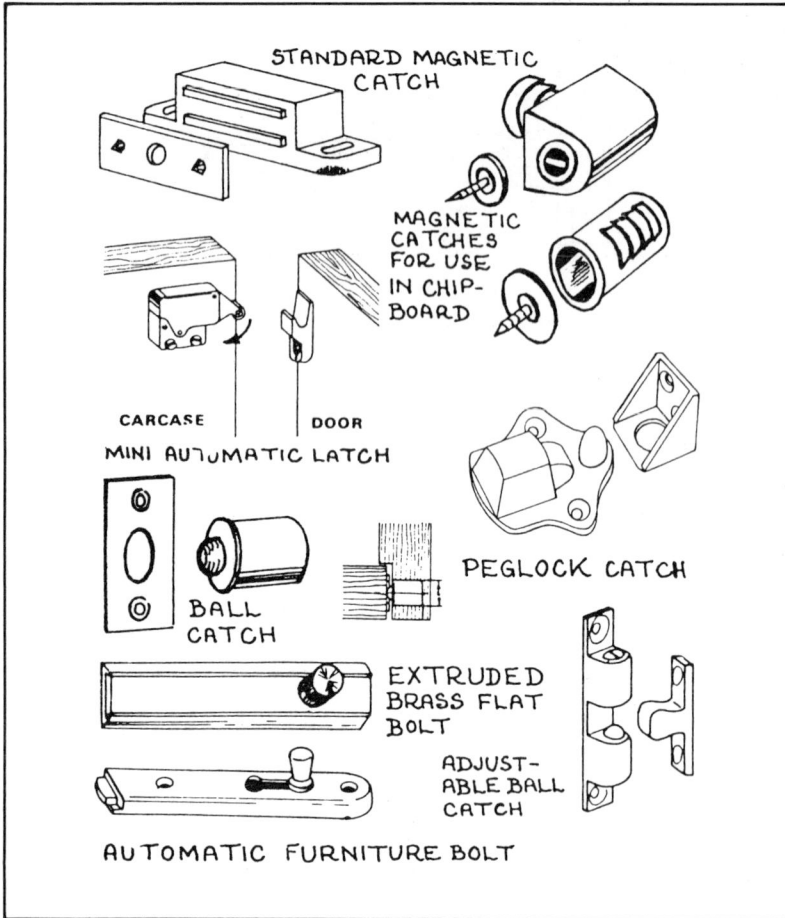

Fig. 200   Catches and bolts

door in a closed position but not to lock it. They are usually a small mechanical device which operates automatically. Catches and bolts are made from steel with a bright zinc or chrome finish, brass, or a tough plastic such as nylon.

Locks

The straight cupboard lock, and the 'cut' locks illustrated (Fig. 201) are top quality brass fittings for traditional cabinet work. The so-called 'cut' locks have to be cut into the door, drawer or box. Modern locks may do the same work, but few match the robust construction or efficiency of traditional locks.

Modern locks (Fig. 202) are however, often easier to fit requiring only one hole and a few screws. In many cases however, they are less adaptable being designed for one specific purpose. The need for a greater range of locks has arisen because many different materials are now used in cabinet work e.g. panel products, glass and perspex.

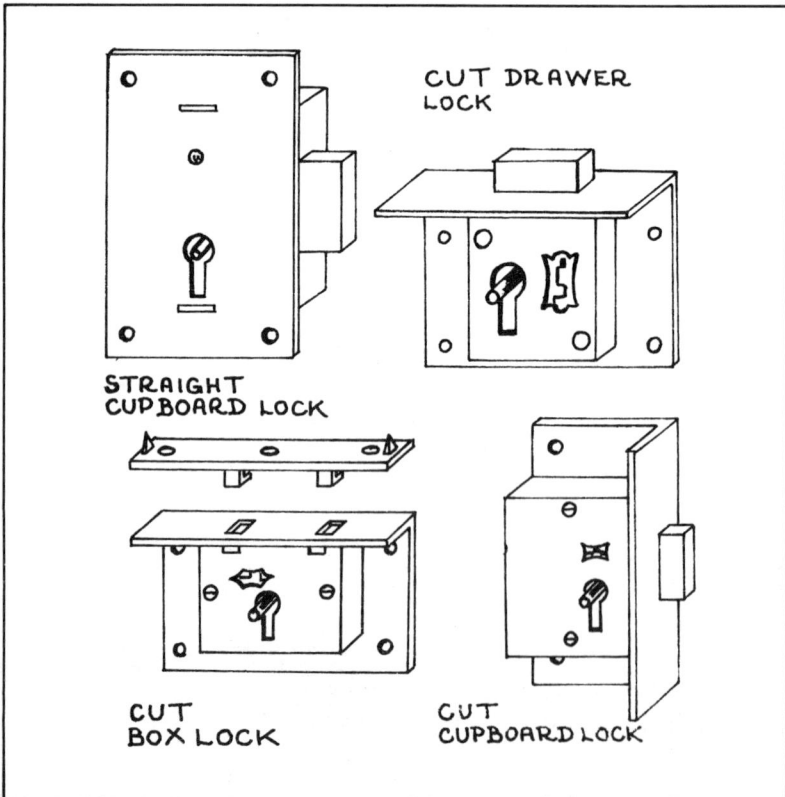

Fig. 201   Traditional locks

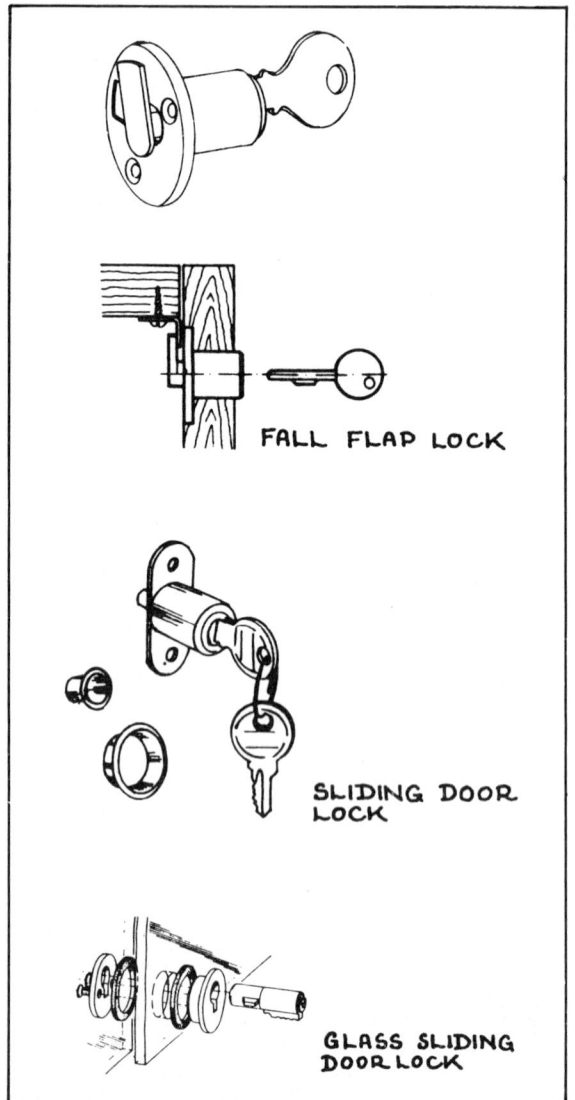

Fig. 202   Modern locks

Knobs and handles

There is a great variety from which to choose (Fig. 203). They are made from natural wood, plastics and metals, often with plated finishes. They may be fixed with glue, screws, back bolts or be edge fitted.

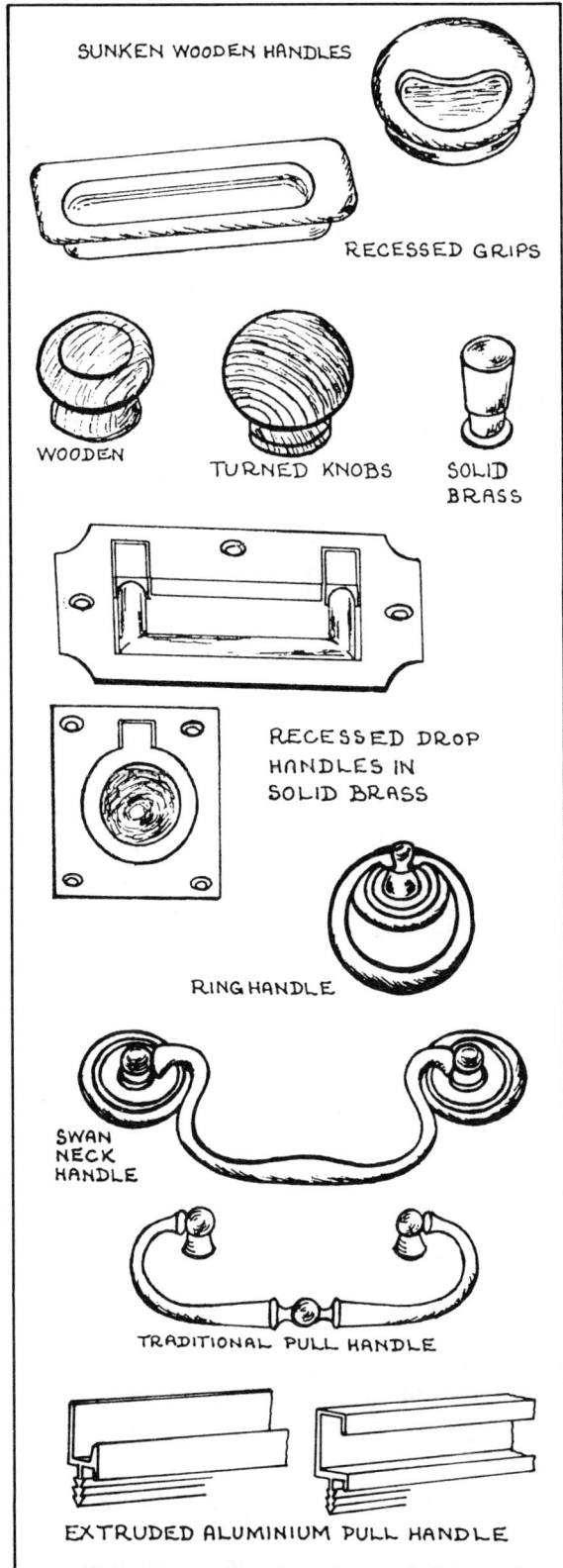

Castors

Castors (Fig. 204) allow furniture to be moved easily without lifting. Some furniture is often quite heavy so castors spread the load evenly on the floor. Therefore they have to be strongly built.

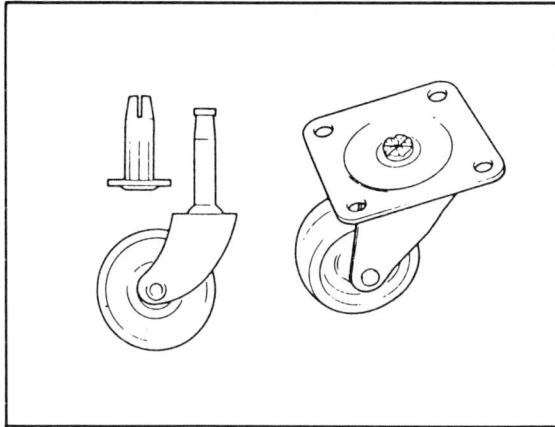

Fig. 204   Castors

Stays

These (Fig. 205) support flaps or limit the opening of lids and doors. Many stays can be used in more than one way so each can have a variety of applications.

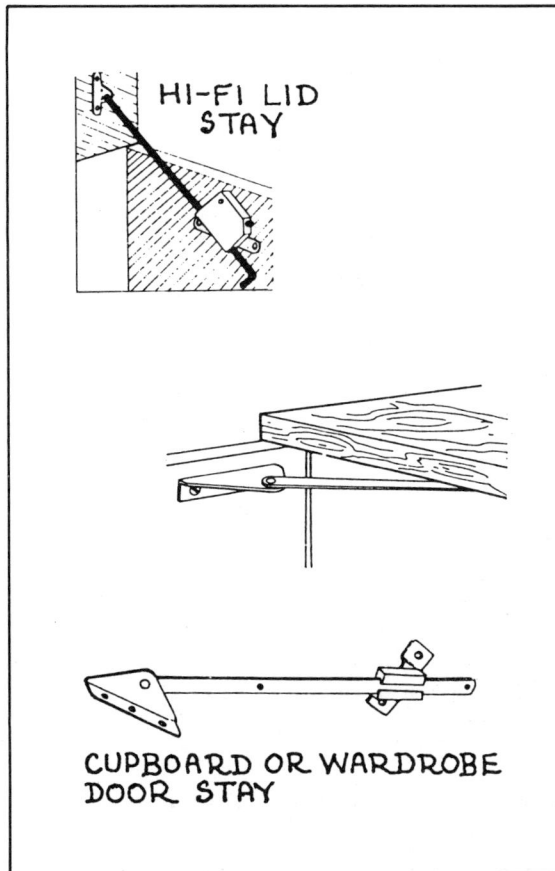

Fig. 203   Knobs and handles

Fig. 205   Some uses of stays

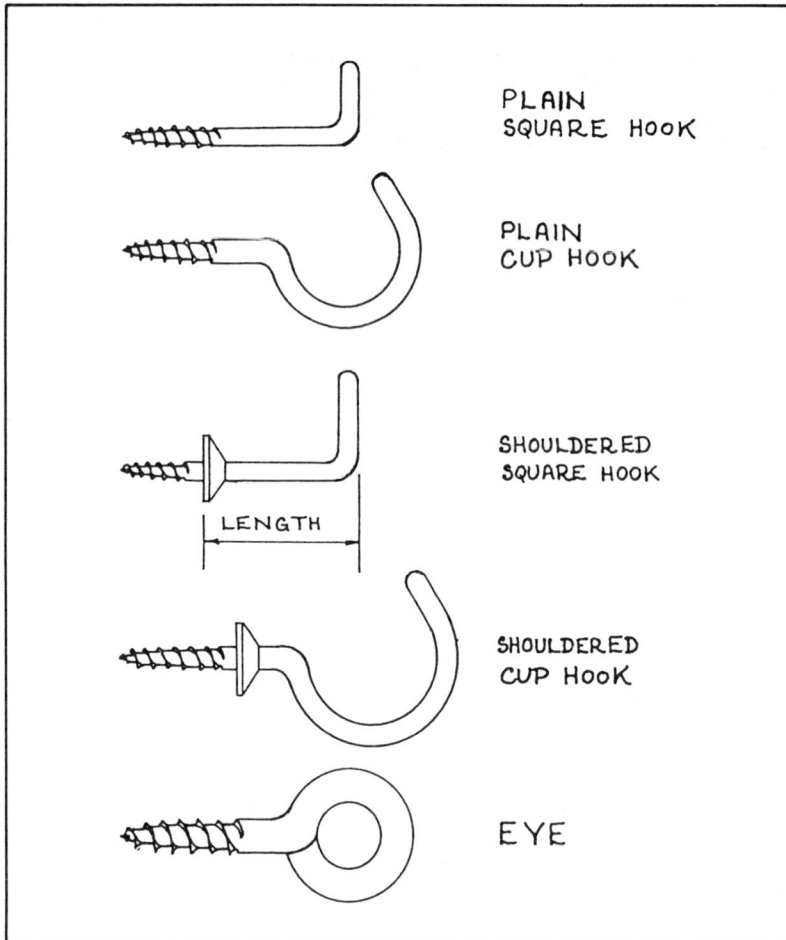

Fig. 206   Screw hooks and eyes

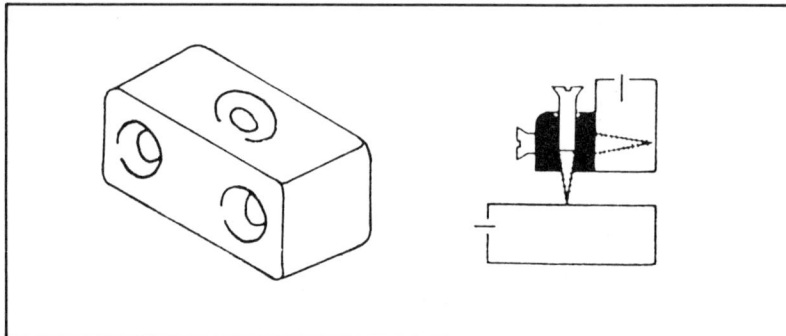

Fig. 207   Modesty 'bloc'

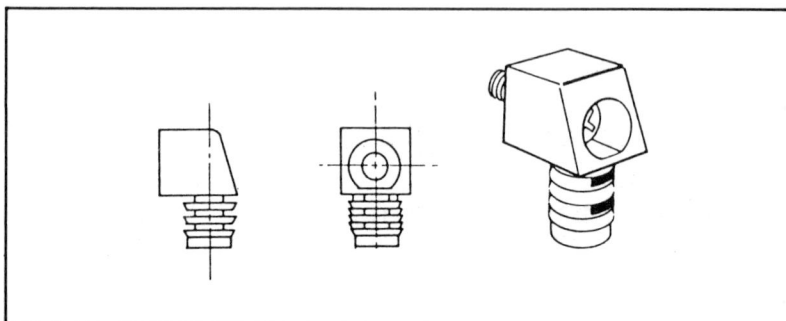

Fig. 208   Mini-'block'

Screw hooks and eyes

These have many uses (Fig. 206). Made from steel, brass, aluminium alloy with plated and sometimes plastic finishes there is a wide range from which to choose.

QUESTIONS

1. Draw 3-dimensional sketches to show the difference between (**a**) a single-ball catch and (**b**) a magnetic catch.

2. A small cabinet has been fitted with double doors. Draw the type of sliding bolt which could be used to secure the left hand door.

3. (**a**) What are the materials from which 'fittings and fixings' are made? (**b**) Some are 'self-coloured'; what other finishes are available?

4. Make sketches from different positions to show the features of the 'cut' cupboard lock.

5. Design a small brass knob to be fitted to the door of an oak cabinet. How would it be fixed?

4.6 'KNOCK DOWN' (KD) FITTINGS

Woodscrews are the simplest form of KD fitting and have been used for many years to fasten sections of furniture together. Until recently very few special fittings were used but during the last few years the greater demand for specialised KD fittings has resulted in a wider range of ingenious fixing devices.

Modern man-made panels such as coated or veneered chipboard, plywood and blockboard have many excellent qualities for furniture construction, but are difficult to joint in the traditional way. KD fittings overcome this disadvantage; in fact KD fittings have made it possible to exploit the strength characteristics of these panel products. They can be fitted together into strongly constructed cabinets and other units without the need for framework.

There is no single all-purpose KD fitting. Several different types may be needed to assemble one work-piece.

Modesty 'bloc'

The Modesty 'bloc' (Fig. 207) is a simple three-screw fitting made in white, brown or beige plastic. It is neat, but being a one-piece fitting is best suited for light cabinet work, shelf supports, and plinth and pelmet construction where there is little stress.

Mini-'bloc'

A simple light-weight fitting (Fig. 208), it requires only two holes for attachment. Deep-threaded

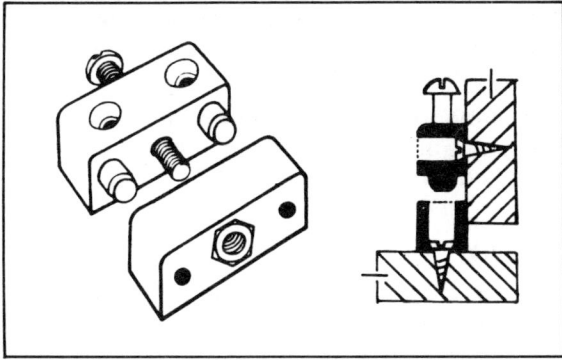

Fig. 209 Bloc-joint fitting

screws with combination slotted/cross heads are used. Their deep thread bites firmly into the wood.

Bloc-joint fitting

This is the most popular of all KD fittings (Fig. 209), made from hardened plastic complete with machine screw and retaining nut. The two moulded guide dowels ensure accurate alignment of the two parts. This provides a stronger joint than the two smaller blocks.

Confirmat panel connecting system

This is a versatile system (Fig. 210) which is very strong and stable, and can be used for framework construction, for connecting panels to a framework, or for connecting panels together. It is simple to fit and although not strictly an invisible fitting the use of cover heads gives it a smooth appearance.

At first sight the Confirmat connector looks like a 'rough threaded machine screw', but it is effective. With a deep cylindrical buttress thread that bites deeply into the chipboard or wood fibre, it provides a much stronger hold than more conventional connectors or screws.

Scan screw fittings

This range of fittings is now the principal jointing medium in frame and panel construction in furniture. Their quality, strength, ease of fitting and attractive finish make them ideal KD fittings for use in the workshop.

A machine screw with a countersunk head is used with a brass **collar** and steel **cross dowel** (Fig. 211). The screw has a hexagonal socket in the head and is turned with an **Allen key**. This collar and dowel combination together with the efficient tightening action of the Allen headed screw produces a close, strong joint without damaging the materials being jointed.

Screw socket

The screw socket (Fig. 212) gives a strong concealed thread in wood, chipboard etc. Drill a Ø 10 mm hole to the depth required and drive in

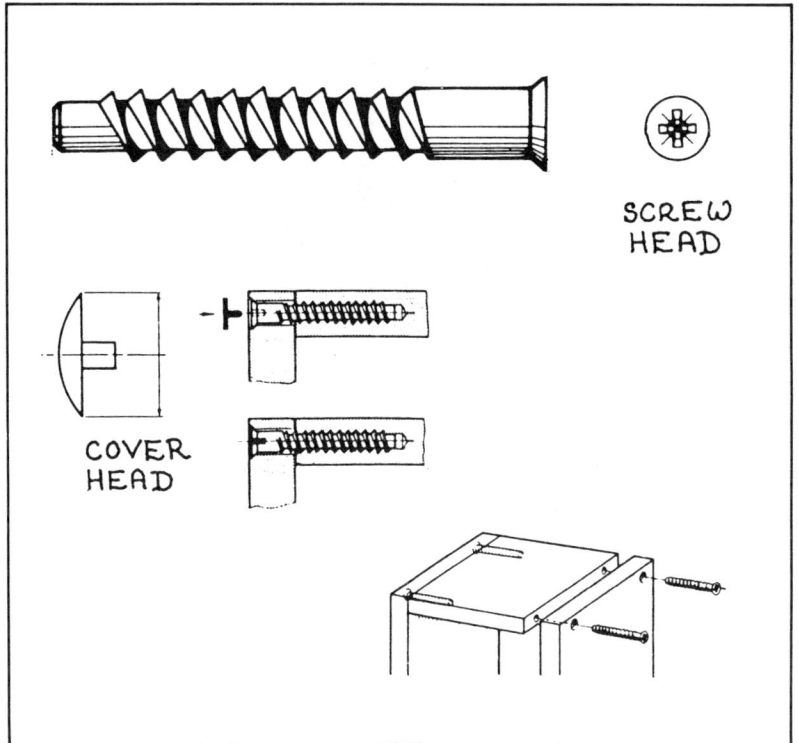

Fig. 210 Assembling with Confirmat connectors

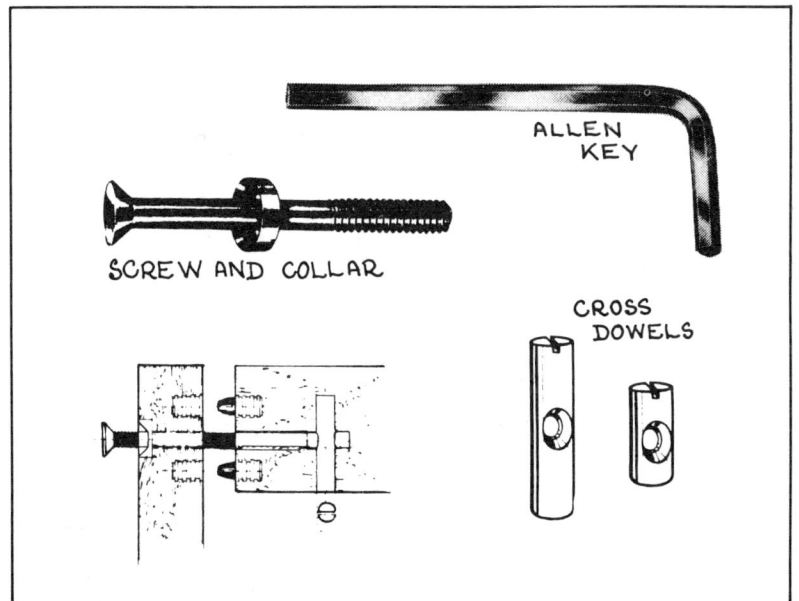

Fig. 211 Scan screw fittings

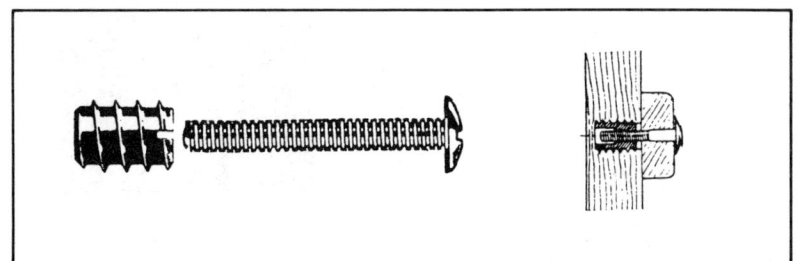

Fig. 212 Screw socket and standard washer head bolt

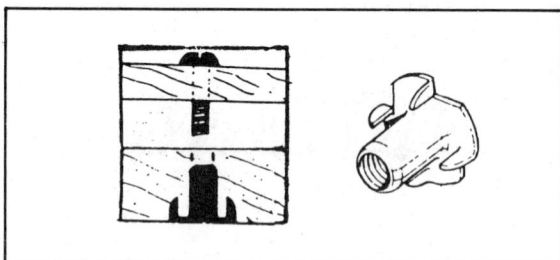

Fig. 213   Tee nut

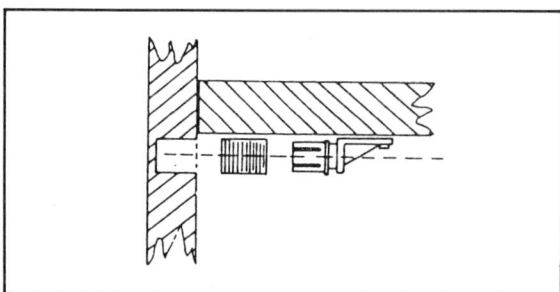

Fig. 214   Plastic shelf studs

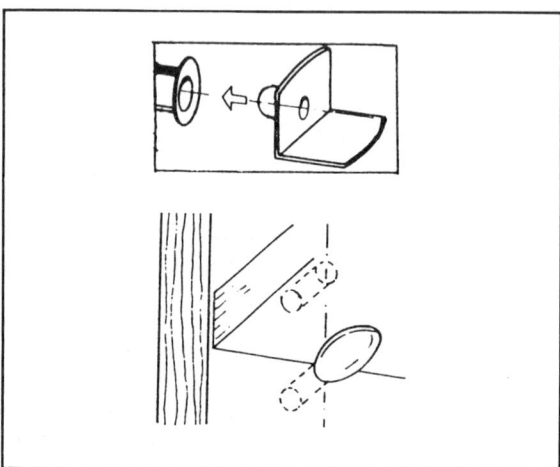

Fig. 215   Metal shelf studs

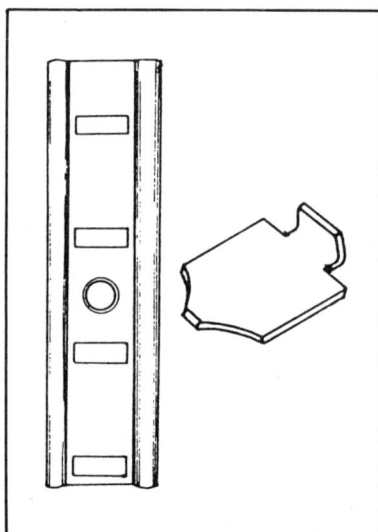

Fig. 216   Book case strip and clip

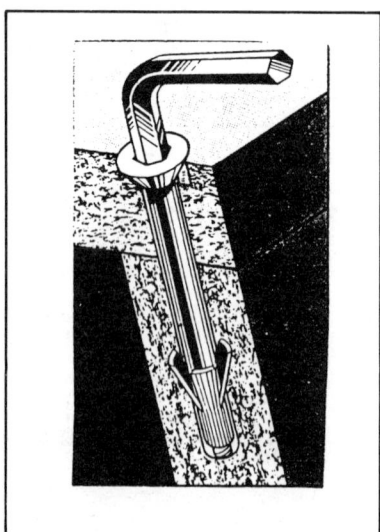

Fig. 217   Harpoon bolt tightened
by an Allen key

the screw socket with a screwdriver. Where wood thickness will permit, fasten up to 3 mm below the surface for greater holding power.

The **standard washer head bolt** (Fig. 212) is an invaluable general purpose fitting with an almost unlimited range of uses.

Tee nuts

Tee nuts (Fig. 213) are inserted from the back of one member of the workpiece, with only the screw hole showing on the face. An 8 mm hole is drilled into the wood and the Tee nut is pressed home, the prongs sinking into the surrounding wood to locate and firmly anchor the nut. A bolt threaded into the reverse side of the Tee nut as illustrated, will serve to further draw the nut into the wood.

Shelf fittings

Plastic shelf studs (Fig. 214) have a small bracing piece below the platform to support the load.

Metal shelf studs (Fig. 215) can support a heavier load than plastic studs. With a nickel or electro-brassed finish, they are fitted in much the same way as the plastic supports.

Bookcase strips (Fig. 216) designed for surface fitting only, need to be screwed to the insides of the bookcase. They are made from steel with bronzed or bright zinc plate finish. **Shelf support clips** locate in the horizontal slots and are in matching finishes.

Harpoon bolt

This is suitable for use in most types of timber and particle boards from 15 mm upwards (Fig. 217). It is ideal for the assembly of KD furniture as the harpoon tip bites deep, locks tight and stays in place if the screw is removed.

QUESTIONS

1. (a) What do you understand by the term 'knock-down'? (b) Which is the simplest KD fitting?

2. Why are KD fittings particularly necessary for use with modern manufactured boards?

3. (a) Sketch the 'modesty bloc' and the 'bloc-joint' fitting, to show clearly how they differ. (b) Which makes the strongest joint and why?

4.7 CLEANING UP

Assembling refers to the final putting together of a piece of work. Depending upon its design and function this may entail the use of nails, screws and adhesives, or a combination of these. In certain circumstances assembly may be solely by means of 'knock down' fittings.

Work to be assembled has to be 'cleaned up' so that it will accept a suitable 'finish'. There are, of course certain exceptions such as rough interior or exterior work, fences, garden woodwork etc.

Cleaning up is usually done in two stages.
1. Inner surfaces and any surfaces which may be difficult to do later must be cleaned up before being assembled.
2. All remaining outer surfaces are then cleaned up when the glue has set and any necessary cramps or holding devices have been removed.

Cleaning up means removing surplus pencil lines left from setting out, dirt and finger marks which may have been left on the wood, and finally smoothing rough grain. This may be around knots, or be crossed or interlocking grain which may have to be scraped with the cabinet scraper. Cleaning up also includes removing 'horns' and any waste wood protruding from joints. This is usually done with the tenon saw and smoothing plane.

The final cleaning up of the work is done with an abrasive paper wrapped around a suitable cork or softwood block, to produce a smooth surface on the wood. This 'sanding' must always be done working 'with the grain'. If the abrasive is rubbed 'across the grain', deep scratches are produced which are very difficult to remove. Even the smallest surface defect is exaggerated and becomes more noticeable when stains or polishes are applied. Therefore the surface preparation must be thorough.

It is good practice when cleaning up to obtain the final smooth surface, to 'raise the grain' by damping the surface with a wet cloth. Then allow it to dry and rub down again with fine abrasive.

When assembled work is being cleaned up it must always be held firmly, either in the vice or cramped to the bench top. This avoids the possibility of 'chatter' marks on the wood, caused by the plane if the work vibrates. Also both hands are free to use the abrasive, or in the case of large work, the orbital sander.

Handtools used for cleaning up are:
1. Smoothing plane (See 1.5) or in the case of very small work or difficult end grain, the block plane (See 1.6)
2. Cabinet scraper (See 1.7)
3. Abrasive papers, or orbital sander (See 1.13)

Abrasives

Several types of grit are used to make abrasive papers. The most common grit is **glass** but although it produces a smooth surface, being comparatively soft, it wears rapidly and soon becomes 'loaded'. **Garnet** is a natural mineral

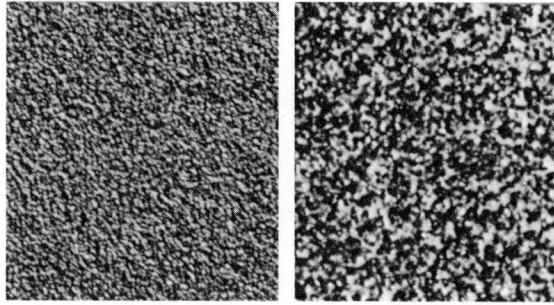

CLOSEKOTE   **OPENKOTE**

Fig. 218

producing sharp individual grits. It is harder than glass, and is therefore longer lasting. **Aluminium oxide** produces sharp abrasive grains. It is hard wearing and gives excellent results on hardwoods, painted surfaces, plastic fillers, metals and many other materials.

'Wet and dry' papers made from **silicon carbide** are used less frequently, but do give excellent results in scuffing old paint, and for between-coat sanding of primers, paints, lacquers and enamels. **Tungsten carbide** grits are even more durable with a life one hundred times that of glass. They give good results on fibre glass and plastic laminates, as well as wood.

The grits are bonded with animal or resin glues to a backing sheet of paper, cloth, or a combination of these, depending on whether it is to be used by hand or machine. **'Closekote'** (Fig. 218) is the standard coating, the whole surface being covered with abrasive grains. **'Openkote'** has 50 to 70% abrasive grain coverage on the backing, leaving open spaces between the grains. This reduces 'loading' or clogging up of the abrasive surface.

Various grades of steel wool, or Bear-tex hand pads are also useful finishing abrasives. These are made from a non-woven backing of nylon fibres bonded with waterproof resins in four grades, from very fine to coarse.

QUESTIONS

1. What does the term 'assembly' mean?

2. **(a)** What is 'cleaning up' and why is it necessary? **(b)** What are the two stages of cleaning up?

3. **(a)** Name the tools used for cleaning up purposes. **(b)** Make sketches of each one.

4. **(a)** What are the grits commonly used to make abrasive papers and why do some abrasives last longer than others? **(b)** How are abrasive papers made?

5. **(a)** What precautions should you take when using abrasive papers. **(b)** What else other than grits can be used as finishing abrasives?

4.8 FINISHING

When 'cleaning up' has been completed the work is ready to receive its 'finish'. All finishes give a degree of protection to the surface of the work and to some extent the wood from which it is made. They protect against dirt and give some protection against moisture. Some finishes also protect against heat. Finishes which protect against insect or fungus attack are called preservatives.

The natural beauty of wood can be enhanced by giving it a clear 'natural finish' which can be high gloss, eggshell or matt. Where necessary stains are used to change the colour or emphasise the grain. Paints are used where solid colour is required and must be applied according to the

Fig. 219   Effects of wood dyes on softwoods

Fig. 220   Dark shade applied to a hardwood veneer

manufacturer's instructions. Paints containing lead should be avoided, particularly for childrens' toys, as they are toxic.

Any nails or panel pins should be punched in and the hole 'stopped' with a matching filler. Similarly, open-grained wood such as oak may have to be filled with a filler. These are rubbed in with a cloth and then sanded off, and may have to be tinted with stain before application to ensure matching colours.

Wood dyes

Wood dyes (Fig. 219) are best applied using a 'rubber' made from a wad of lint free cloth, which is rubbed up and down with the grain. Care should be taken not to overlap the dye to avoid streaking. Over application should be avoided. A second coat can be applied if a darker finish is required, but if too much is applied in the first place it cannot be made lighter. All dyes are transparent and light fast, and all can be intermixed to give an almost infinite choice of wood shades or colours. Some hardwoods and resinous softwoods may not accept the deep penetration of dyes, in which case polyurethane stained varnish wood shades may be used as an alternative.

Wood treated with wood dyes may be finished with clear polyurethane varnish, French polish, or good quality bees wax. For outdoor work exterior varnish will give effective protection. As the application of a finish will change the shade of a wood dye it is advisable to experiment with an off-cut of wood to ensure getting the correct colour.

Polyurethane coloured matt stains

These modern stains give a colourful finish to modern woodwork. They are intended for interior use only, and contain transparent dyes which allow the grain to show through. The stains are best used on light coloured woods, then sealed with a coat of clear varnish in either gloss, eggshell or matt, depending upon the surface finish required.

Stained-varnish wood shades

These are a much improved version (Fig. 220) of the old-style varnish stains. The varnishes are applied by brush, and stain and varnish in one operation. Further coats can be applied if required to get a darker shade. If necessary the whole surface can be finished with a clear varnish. The wood should be rubbed down lightly between coats with fine or very worn glasspaper.

Clear polyurethane varnish

Most interior or exterior timbers can be varnished, bringing out the full natural beauty of the grain and giving a long lasting protective finish. Polyurethane varnish is one of the toughest and most durable of clear finishes, suitable for use on kitchen surfaces, work tops, furniture, floors, doors, etc. It is resistant to wear, knocks, liquids, dilute acids and limited heat. Application should be with a fine bristled brush, with careful rubbing down between coats. If eggshell or matt finishes are to be applied the initial 'build up' should be with high gloss.

Exterior varnish

Exterior varnish must be flexible to withstand the expansion and contraction of wood caused by changing atmospheric conditions. Failure to do this can result in a breakdown of the protective coat, allowing water to get behind it, resulting in discolouration and flaking.

Super yacht varnish is exceptionally durable and will retain its high gloss for many years. It is moisture repellent and will withstand extreme weather conditions. Both it and exterior varnish resist the effects of ultraviolet sunlight, which changes the natural colour of wood.

Varnishing techniques

Preparation of the surface is most important. New wood must be sanded with the grain and finished off with a fine or flour grade paper. Surfaces painted or varnished previously may have to be stripped with a paint stripper.

Application

Apply a good coat to the surface, following the grain of the wood. Cross brush with long strokes to even the film, and 'lay off' with the tips of the bristles, following the grain. Sand with fine or flour grade paper between coats. In cross brushing there may be an accumulation of varnish at the edges but this can be removed by lightly brushing with an almost dry brush.

On mouldings and vertical surfaces 'runs' may occur. The work should be examined after 10 to 15 mins and if 'runs' have occurred again brush out with the tip of an almost dry brush.

The surface of timber which has been stripped in caustic soda or with alkaline paint removers should be neutralised; this can be done with vinegar, followed by subsequent washing and glass papering. New woodwork may be wiped over with white spirit to ensure the surface is clean, dry and free of wax, oil or grease.

French polish

French polishing (Fig. 221) is one of the traditional finishes used by craftsmen. Common shades are Button (yellow), French (paler) and White. Your choice depends upon the colour of the wood being polished and the tone you wish to achieve. Considerable practice is necessary to acquire the skill to produce the thin transparent and glass-like surface.

Brush polishing is very similar and should be applied rapidly with a soft brush to get a good high gloss French polish effect.

Teak oil

This is manufactured from synthetic resins and oils. It will preserve the natural beauty of teak, afromosia and all fine quality furniture where a matt oiled finish is required. It is also ideal for garden furniture, wall cladding, doors, veneered chipboard etc. Normally only a thin coat should be applied with a soft rag, but for exterior use brush on a more generous coat and allow to dry for a day.

Teak oil is combustible and oily rags should be destroyed immediately after use. Keep away from naked lights.

Wax

Bees wax is still a popular finish which gives a mellow glow to the work. 'Chilled' wax now gives a hard shine in much less time. Application and polishing are with a soft cloth.

Paints

Work requiring a 'solid' coloured finish has to be painted. For small work enamel paint is recommended because it is quick drying, the colours are dense, usually needing only one coat, and it is generally available in small quantities.

For the best results apply a coat of white or pink primer depending on the colour of the final coat. The primer seals the wood. Then finish with the appropriate undercoat and enamel. On hard and resinous woods aluminium primer is recommended. For large work it is advisable to use standard gloss or satin finish paints.

Application

As for varnish, the surface must be clean, dry and free from dust. The paint must be applied with a clean fine bristle brush.

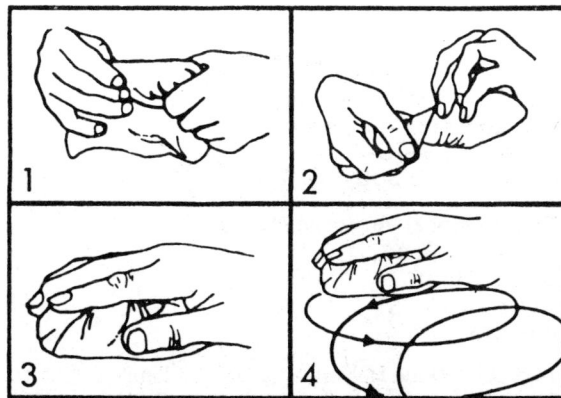

Fig. 221 Applying French polish with a 'rubber' made from cloth

QUESTIONS

1. In what ways do 'finishes' protect woodwork?

2. (a) When would you use a 'filler'? (b) How is this done?

3. (a) Describe the use of wood dyes. (b) How can you ensure you obtain the required colour?

4. (a) Which finishes can be applied over wood dyes? (b) Which is most suitable for outdoor work and why?

5. Write clear instructions for (a) the surface preparation of wood (b) the correct application of clear polyurethane varnishes.

REVISION QUESTIONS

Nails and screws

1. Contrast the uses of the following nails (a) round wire (b) panel pin (c) escutcheon pin (d) upholstery nail (e) cut tack. Illustrate your answers with sketches.

2. Draw full-size a 4" (100 mm) no. 12 counter-sunk slotted head woodscrew with a gimlet point. Remember that this type of screw is threaded for $\frac{2}{3}$ of its length.

3. Nails, screws and bolts are mechanical ways of fastening wood together. Make simple sketches, with notes to show when you would use each one.

4. What things should you consider when selecting a screw for a particular piece of work?

5. (a) From which metals are screws made? (b) When ordering screws how are they specified?

6. Give step-by-step instructions on how to fit an inset screw cup.

Adhesives

7. List and describe the different types of glue used in school workshops.

8. (a) How would you prepare Scotch glue? (b) For what kind of work is it used and why?

9. Design a piece of work which because of its use or position should be made with waterproof adhesive.

Hinges

10. Using step-by-step sketches show how you would 'box in' one of a pair of small brass butt hinges on a jewellery box.

11. Using a labelled sketch show how you would fix a cupboard stay to a cupboard door.

12. Design a simple metal stay which will hold a pair of steps in the open position and fold when they are closed.

'Knock down' fittings

13. Design a small open-fronted bookcase with adjustable shelves the carcase of which is to be made from veneered chipboard and assembled entirely with 'modesty bloc' and 'bloc-joint' fittings. Your design should include a back, and toe recess with 'kicking panel'. Show clearly how these would be fitted and fixed.

14. Make a sectional drawing to show how the rail of a small modern table could be jointed to the leg using Scan screw fittings.

Cleaning up and finishing

15. A piece of mahogany has interlocking grain. How would you clean it up before applying a finish of clear varnish?

16. You have a small box to assemble and clean up. (a) List the stages involved. (b) List all the tools and materials needed to prepare it for 'finishing'.

17. (a) Where would you use an enamel instead of a paint, and in what ways do they differ? (b) Should the use of any particular type of paint be avoided and why?

18. (a) How would you prepare a piece of work for an enamelled or painted finish? (b) Which paint is specially recommended for hard or resinous woods?

19. Which polishes are applied with a 'rubber' or a rag? Write short notes about each.

20. Explain the difference between, and give examples of the type of work on which coloured matt stains and stained varnish wood shades could be used.

© T. Pettit 1986

First published in Great Britain 1986 by
Edward Arnold (Publishers) Ltd
41 Bedford Square
London WC1B 3DQ

Edward Arnold (Australia) Pty Ltd
80 Waverly Road
Caulfield East
Victoria 3145
Australia

Text set in Neue Helvetica by Oxprint Limited, Oxford.
Printed and bound in Great Britain by Bath Press Ltd,
Bath, Avon and bound by W. H. Warnes & Sons Ltd,
Clevedon, Avon.

British Library Cataloguing in Publication Data

Pettit, T.
　Woodwork technology
　1. Woodwork—Examinations, questions etc
　I. Title
　684'.08'076　　　　TT185

　ISBN 0-7131-8281-4